Vortex Structures in a Stratified Fluid

APPLIED MATHEMATICS AND MATHEMATICAL COMPUTATION

Editors

R.J. Knops, K.W. Morton

Text and monographs at graduate and research level covering a wide variety of topics of current research interest in modern and traditional applied mathematics, in numerical analysis, and computation.

1 Introduction to the Thermodynamics of Solids *J.L. Ericksen* (1991)

2 Order Stars *A. Iserles and S.P. Norsett* (1991)

3 Material Inhomogeneities in Elasticity *G. Maugin* (1993)

4 Bivectors and Waves in Mechanics and Optics *Ph. Boulanger and M. Hayes* (1993)

5 Mathematical Modelling of Inelastic Deformation *J.F. Besseling and E. van der Geissen* (1993)

6 Vortex Structures in a Stratified Fluid; Order from chaos *Sergey I. Voropayev and Yakov D. Afanasyev*

7 Numerical Hamiltonian Problems *J.M. Sanz-Serna and M.P. Calvo*

(Full details concerning this series, and more information on titles in preparation are available from the publisher.)

Plate 1 A starting vortex quadrupole in a stratified fluid, viewed from above (thymol blue visualization).

Plate 2 Coherent dipolar vortices in decaying turbulence in a stratified fluid, viewed from above (thymol blue visualization).

Vortex Structures in a Stratified Fluid

Order from chaos

SERGEY I. VOROPAYEV

and

YAKOV D. AFANASYEV

CHAPMAN & HALL

London · Glasgow · New York · Tokyo · Melbourne · Madras

Published by Chapman & Hall, 2–6 Boundary Row, London SE1 8HN, UK

Chapman & Hall, 2–6 Boundary Row, London SE1 8HN, UK

Blackie Academic & Professional, Wester Cleddens Road, Bishopbriggs, Glasgow G64 2NZ, UK

Chapman & Hall Inc., One Penn Plaza, 41st Floor, New York NY 10119, USA

Chapman & Hall Japan, Thomson Publishing Japan, Hirakawacho Nemoto Building, 6F, 1-7-11 Hirakawa-cho, Chiyoda-ku, Tokyo 102, Japan

Chapman & Hall Australia, Thomas Nelson Australia, 102 Dodds Street, South Melbourne, Victoria 3205, Australia

Chapman & Hall India, R. Seshadri, 32 Second Main Road, CIT East, Madras 600 035, India

First edition 1994

Typeset in 10/12 Times by Thomson Press (India) Ltd, New Delhi, India
Printed in Great Britain by the University Press, Cambridge

ISBN 0 412 40560 1

A catalogue record for this book is available from the British Library

Library of Congress Cataloging-in-Publication data available

∞ Printed on permanent acid-free text paper, manufactured in accordance with ANSI/NISO Z39.48–1992 and ANSI Z39.48–1984 (Permanence of Paper).

Contents

Preface

Large-scale vortex structures are widespread in nature and they play a crucial role in geophysical and industrial chaotic flows. These ordered structures are easily detectable and are visible to the naked eye in many quasi-two-dimensional liquid systems, in particular, in the stratified ocean. The human eye possesses the unique ability to store and compare pictures of a flow at different times and it automatically carries out the complicated operations connected with this procedure (including expansion of the coordinate system). In many cases a sequence of two or three images is enough to allow the preliminary conclusion that the images can be considered to be similar at different moments in time. For this reason numerous pictures of the flows under study are included and widely used in this book. The idea of self-similarity (where possible, verified experimentally) essentially simplifies the subsequent analysis and in many cases general mathematical models can be developed.

This book is devoted primarily to the consideration of self-similar structures arising in quasi-two-dimensional liquid systems under the action of external forces. The interactions of such structures with each other and solid boundaries are studied, and particular attention is paid to the physical aspects of the problem. Because viscosity plays an important role here, the interpretation and analyses are based on a general consideration of the dynamics of vorticity in a viscous fluid. To make the book compact, the fundamentals of fluid dynamics are included in a compressed form.

This book is based on a succession of notes prepared for a graduate course that the authors are teaching in the Department of Ocean Thermohydromechanics at Moscow Physical-Technical Institute. Naturally, the text draws heavily on our own research and that of our students. We hope that the book will be of interest to specialists in fluid dynamics, geophysical fluid dynamics and applied mathematics. It may also be considered as a teaching aid for graduate and postgraduate students in universities and physical-technical institutes.

The authors are grateful to Igor A. Filippov for his valuable help and constant technical assistance. Our thanks are also due to M.H. Rozovskii and P.G. Potylitsin for numerical calculations.

Sergey I. Voropayev
Yakov D. Afanasyev
Institute of Oceanology, Moscow

Introduction and some geophysical examples

The emergence of coherent vortex structures is a characteristic feature of two-dimensional turbulence and irregular vortical flows. These structures are relevant to large-scale, quasi-geostrophic, quasi-planar geophysical flows. Vortices are abundant in the ocean and atmosphere. With the advent of satellites, it has become clear that large-scale vortices are present in virtually all parts of the ocean. Because they can effectively transport momentum, heat, salt, biochemical and artificial products, they play an essential role in ocean dynamics, determining the instantaneous fields of velocity, temperature and salinity. Thus they can be regarded as forming the internal oceanic weather (a term proposed by Kamenkovich, Koshlyakov and Monin, 1987).

Various physical processes cause the motion of water in the ocean. Motions driven by different processes have different spatial and temporal scales. The lower limit of the spatial scale ($L \approx 10^{-1}$ cm) is determined by the process of smoothing of sharp gradients by molecular transport of momentum, heat and salt. The upper limit ($L \approx 10^{9}$ cm) is determined by the dimension of the ocean. Thus motions over a huge range (ten orders of magnitude in length-scale variations) can exist together in the ocean. Oceanographers classify these motions into the following general groups (Kamenkovich, Koshlyakov and Monin, 1987):

- small-scale motions ($L = 10^{-1}$–10^{4} cm), which include in particular quasi-isotropic turbulence and wave motion from capillary waves to surface gravity waves and short internal waves;
- mesoscale motions ($L = 10^{4}$–10^{6} cm), which include long internal waves, internal oscillations and tidal currents in shallow waters;

- synoptic motions ($L = 10^6 - 10^7$ cm), which include frontal and free oceanic vortices;
- planetary motions ($L = 10^7 - 10^9$ cm), which include large currents crossing the ocean.

Although in general the motions of water in the ocean can be described as 'turbulent', there are a number of specific features arising on the widely different length scales that occur. The physical factors governing particular flow regimes are of great importance. For example, molecular transport is relevant for small-scale quasi-isotropic motion, buoyancy forces due to stratification play an important role for larger-scale motion, very large-scale motion is affected by the rotation of the Earth, while the variation of the Coriolis parameter with latitude (the so-called β-effect) plays a crucial role in planetary-scale motion. Observations of ocean dynamics demonstrate that the mean intensities of motions on different scales vary greatly. As a result, the kinetic energy spectrum of oceanic motion has a minimum for mesoscale motion (Woods, 1980). This minimum separates the energy-containing ranges of small-scale and synoptic motions. This spectral gap shows that large-scale motions exist on a background field of small-scale motions. The idea of separation of length scales appears to be very fruitful, providing the basis for the existence of so-called coherent vortex structures, i.e. large-scale ordered motions in the field of chaotic vorticity fluctuations creating the background turbulence. Direct determinations of the coefficient of horizontal turbulent exchange, obtained from long-term empirical spectra of kinetic turbulent energy in the ocean, give values in the range $\nu_* = 10^5 - 10^8\ \mathrm{cm^2\,s^{-1}}$ (Monin and Ozmidov, 1981). Thus it is clear that the effective Reynolds numbers for large-scale ordered oceanic vortices are not very large, since the effective viscosity due to the background vortex motion is so great. This is very important, because it gives assurance that vortex structures reproduced at moderate Reynolds numbers in laboratory experiments correctly reproduce, at least qualitatively, the main features of coherent oceanic motions (Barenblatt, Voropayev and Filippov, 1989). Theoretical considerations (Voropayev and Afanasyev, 1991) also predict that another parameter governing the dynamics of vortex structures in a stratified fluid, namely the Richardson number (which is equal to the ratio of buoyancy and inertial forces), becomes asymptotically large and can be neglected. This result makes the above assurance yet stronger.

During the last decade the dynamics of coherent vortex structures as are common in geophysical flows has received considerable attention. Theoretical, numerical and experimental studies have revealed the existence of a number of different vortex types. The most frequently occurring structure is the vortex monopole, consisting of a single set of closed streamlines with a common centre. A variety of examples of oceanic vortices of this type are presented in Robinson (1983).

In the last few years examples of a new form of mesoscale-synoptic (i.e. with horizontal scale 10–100 km) vortex structure, termed a mushroom-like current by Ginzburg and Fedorov (1984), have been observed in great numbers in visual and infrared satellite images of the upper ocean.

A typical mushroom-like current is a jet flow with a system of two vortices of opposite sign at its front (Figure 0.1). It resembles a sliced mushroom – hence the name. Numerous observations (Vastano and Bernstein, 1984; Munk, Scully-Power and Zachariasen, 1987; Stockton and Lutjeharms, 1988; Fedorov and Ginzburg, 1988; Sheres and Kenyon, 1989) have demonstrated that mushroom-like currents are widespread in the ocean. For example, Ahlnas, Royer and George (1987) determined 17 mushroom-like currents in just one of their images. To make the currents visible, some indicator is always needed. In the ocean this is usually broken ice, flowering plankton, sediments or temperature contrast. If the whole ocean could be visualized in some way, we should see a vast number of these structures.

One of the first examples of a mushroom-like current was reproduced in laboratory experiments by Voropayev (1983) (see Figure 5.2 below), and later a similar current was observed by Ginzburg and Fedorov (1984) in the Tartar Strait in spring, when the intense solar radiation melts the ice near the shore. A satellite image of this current is shown in Figure 0.1. The fresh water is less dense than the salty sea water, and spreads on the surface. The density front of the spreading water is unstable, and this results in the formation of mushroom-like currents.

Another oceanic phenomenon, so-called upwelling, sometimes produces similar currents. This upwelling is caused by wind blowing

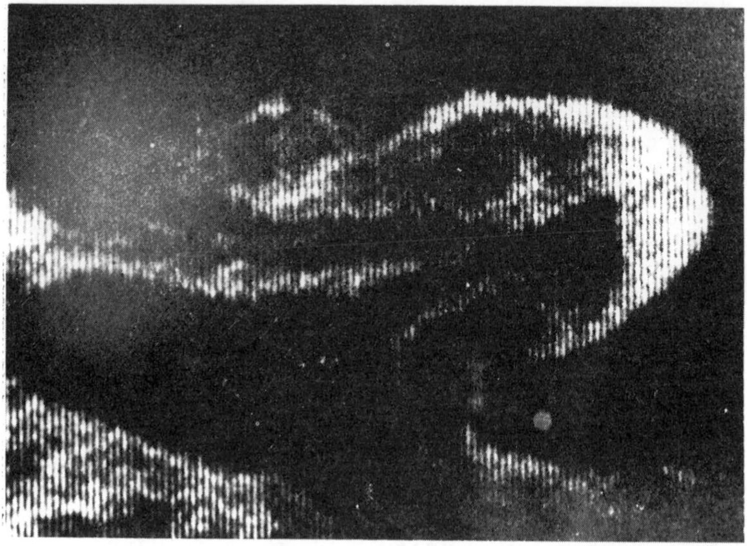

Figure 0.1 *Typical satellite image of a mushroom-like current. The length scale of the flow* ($L \approx 20\,km$) *increases with time, while the propagation velocity* ($U \approx 10\,cm\,s^{-1}$) *decreases. The lifetime is about one to two weeks. (Courtesy of K.N. Fedorov and A.I. Ginzburg.)*

along the shore. Under the action of the Coriolis force, water driven by the wind moves to the right in the Northern hemisphere and to the left in the Southern hemisphere. When the wind blows along the shore so that the resulting motion of surface water is directed off-shore, surface water is replaced by water from below (Figure 0.2). The upwelling water is usually rich in biogenetic elements. This causes rapid growth of living species and the production of chlorophyll, which gives a visible colour contrast in images. The upwelling water is also usually colder than the surface water, but in some cases it is less salty. Heating by intense solar radiation decreases the density of this water, so it spreads on the surface. The spreading water front becomes unstable, and, as in the previous case, a system of transverse jets appears. The typical distance between jets is 50–400 km and their lengths can be as high as hundreds of kilometres, while their depths are restricted to a few tens of metres. Thus the flow is almost planar, and we can expect the formation of numerous mushroom-like currents (Fedorov and Ginzburg, 1988).

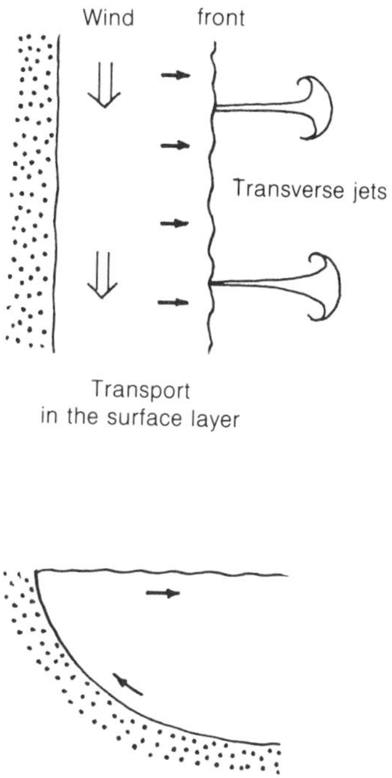

Figure 0.2 *Flow in the upwelling zone (according to Drake et al., 1978) and the formation of mushroom-like currents.*

Another example of mushroom-like currents appears in the formation of rip currents in a surf zone. When waves come into shallow water, they drive the water towards the shore (Figure 0.3). As a result, a strong current forms along the shore. This current separates from the shore and forms narrow jets directed off-shore. These jets have typical 'hats' on their leading edges, and are called rip currents (Drake *et al.*, 1978; Massel, 1989).

One might expect mushroom-like currents to emerge not only at the surface but also in the depths of the stratified ocean. Recent studies by the P.P. Shirshov Institute of Oceanology have confirmed this hypothesis. In 1988 an expedition on the research vessel *Vityaz* studied the formation of eddies in the intrusive lens of Mediterranean water that flows impulsively from the Straits of Gibraltar into the Atlantic. This water is salty and warm, because of intense heating

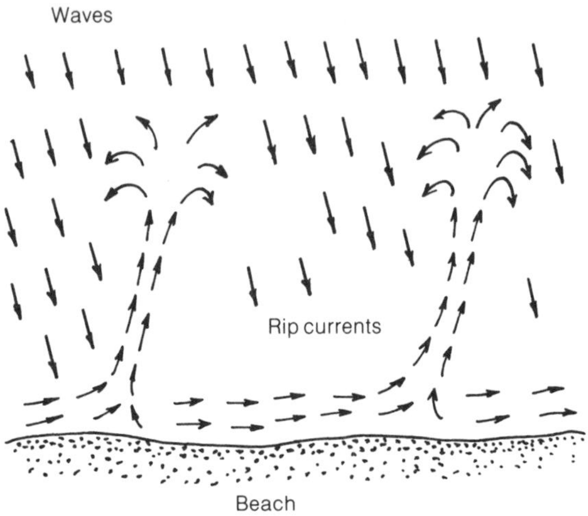

Figure 0.3 *Formation of rip currents in the surf zone (according to Drake et al., 1978).*

and subsequent evaporation in the Mediterranean Sea, and it flows down in narrow underwater canyons of the Cadiz Bay until it reaches the level of equilibrium density. The water mass then intrudes horizontally into the Atlantic, forming a lens of typical horizontal scale of order 100 km and vertical scale of order 200 m. A strong horizontal jet in a stratified fluid must transform into a mushroom-like current, and indeed measurements made in this region have revealed two distinct vortices rotating in opposite directions. Thus, at least initially, the flow is certainly a mushroom-like current. Later the Coriolis force and other factors lead to intensification of anti-cyclonic rotation in the lens and suppression of cyclonic rotation, so that the lens transforms into a circularly symmetric monopolar vortex.

The basic vortex structures mentioned above are shown schematically in Figure 0.4. The simplest vortex monopole is a circular patch of vorticity of one sign. Most vortex monopoles in the ocean consist of a core with vorticity of one sign surrounded by a ring with vorticity of the opposite sign. Thus a vortex monopole possesses axial symmetry (Figure 0.4a). Vortex structures consisting of patches

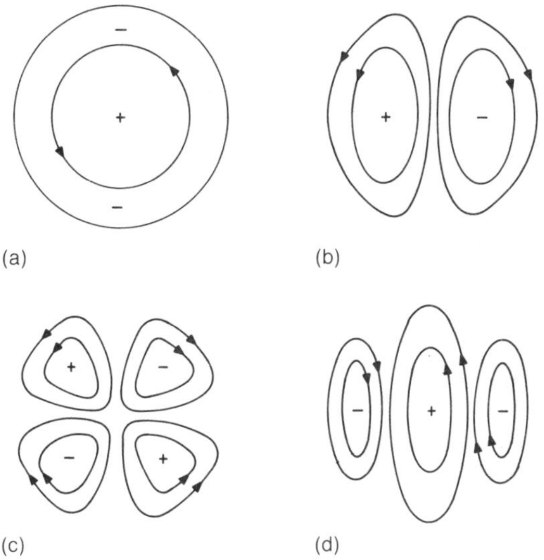

Figure 0.4 *Basic vortex structures: (a) vortex monopole; (b) vortex dipole; (c) symmetrical vortex quadrupole; (d) vortex quadrupole with coinciding centres of positive vorticity (vortex tripole).*

of opposite vorticity can also have other symmetries. A typical example is the vortex dipole, which consists of two patches of opposite vorticity and is characterized by mirror symmetry relative to the plane separating the patches (Figure 0.4b). One can imagine more complex vortex structures. For example, the structure with two symmetry planes shown schematically in Figure 0.4c can be termed a vortex quadrupole. Indeed, results of laboratory experiments demonstrate that symmetric collision of two dipoles produces a vortex quadrupole (Afanasyev, Voropayev and Filippov, 1988; van Heijst and Flor, 1989). A similar vortex quadrupole can appear as the result of disintegration of the unstable vortex monopole of the type shown in Figure 0.4a (Griffiths and Linden, 1981; Kloosterziel and van Heijst, 1991). The asymmetric collision of two dipoles (Orlandi and van Heijst, 1992) or another type of instability of the vortex monopole (van Heijst and Kloosterziel, 1989) causes the creation of another kind of quadrupole (Figure 0.4d) consisting of two dipoles with one common vortex. This type of rotating vortex quadrupole is also called a tripole. One can continue this exercise and introduce formally more and more complex vortex structures, but the question then arises as to the physical origins of such coherent motions in liquid systems.

In general, the stratified ocean, where gravity suppresses vertical motion, is a typical geophysical example of a nearly two-dimensional system where various natural factors act as forcing mechanisms generating vortex motions on different scales. Since each type of vortex is characterized by some basic physical property, an appropriate forcing is needed to generate particular vortex structures. A vortex monopole possesses net angular momentum, and hence some source of angular momentum is needed to generate one. A vortex dipole can be considered as the simplest vortex structure that is characterized by net linear momentum. It is clear that for a volume of fluid to acquire momentum, some force must be applied to it. Thus, in view of their non-zero linear momentum, the generation of dipoles requires the action of a force on a fluid. More complicated vortex structures (e.g. vortex quadrupoles) are created when more complex systems of forces act on a fluid. Thus, in its archetypal form, each vortex structure is generated by the action of a certain forcing, which imparts to the fluid an appropriate physical property.

Besides forcing, another essential condition is required for the generation of stable vortex structures: the flow must be planar. In general, it is not important what physical mechanism suppresses the motion in one direction and thus makes the flow quasi-two-dimensional. For example, vortex dipoles have been observed to occur in a variety of situations:

- in a rotating homogeneous fluid (Flierl, Stern and Whitehead, 1983), where the Coriolis force makes the flow uniform in the direction of the axis of rotation (Greenspan, 1968);
- in magnetohydrodynamic flow (Nguyen Duc and Sommeria, 1988), where the same role is played by the magnetic field;
- in a soap film (Couder and Basdevant, 1986), where the surface tension force makes the film very thin, thus eliminating motion in one direction;
- in a density-stratified fluid (Voropayev and Filippov, 1985), where the gravitational force suppresses vertical motion;
- in a narrow vertical slot, filled with initially stably stratified fluid

and then overturned (Voropayev, Afanasyev and van Heijst, 1993), where numerous dipoles explode in the bulk of fluid (Figure 0.5).

The structures mentioned above can be considered as basic vortex structures in quasi-two-dimensional irregular, turbulent flows. Here the dipoles play a special role. It is a dipole that has momentum. Moreover, isolated vortex monopoles cannot remain single asymptotically in a decaying turbulent flow (Manakov and Schur, 1983). Isolated vortices of opposite signs form dipoles with zero total circulation. Taking this into account, and also in view of their interaction properties, dipoles have been termed the 'elementary particles' of quasi-two-dimensional turbulence. Vortex dipole interaction leads to the emergence of more complex flows. A typical example is a vortex quadrupole, which can be considered as a result of the collision of two dipoles. It seems that vortex dipoles and their combinations are the universal products of any forcing in two-dimensional liquid systems.

Thus it is clear that different forcings generate different vortex structures. However, the important question remains as to the physical property of these flows that allows one to describe them as coherent, ordered structures. Intuitively, these structures are associated with the appearance of order in a chaotic flow. It turns out that a physical property that helps to distinguish ordered motion on a background field of chaotic fluctuations is the fact that this motion conserves its form and structure, i.e. so-called self-similarity. A stricter definition of self-similarity is as follows: a phenomenon developing in time is called self-similar if the spatial distributions of its properties at different times can be obtained from each other by similarity transformations. Although the nature of the self-similarity of coherent vortex structures is not yet completely clear (Cantwell, 1978), this important property plays an essential role in theoretical analysis as well as in the interpretation of experimental results.

Recent studies based on topological considerations (Kiehn, 1992) have demonstrated a strong connection between coherent structures and topological properties of flows of different Pfaff dimension. On the basis of this connection, it was suggested that a coherent structure in a fluid be defined as a compact connected deformable domain of invariant topological properties embedded in perhaps an open, or non-compact, domain of different topology (Kiehn, 1990). Although

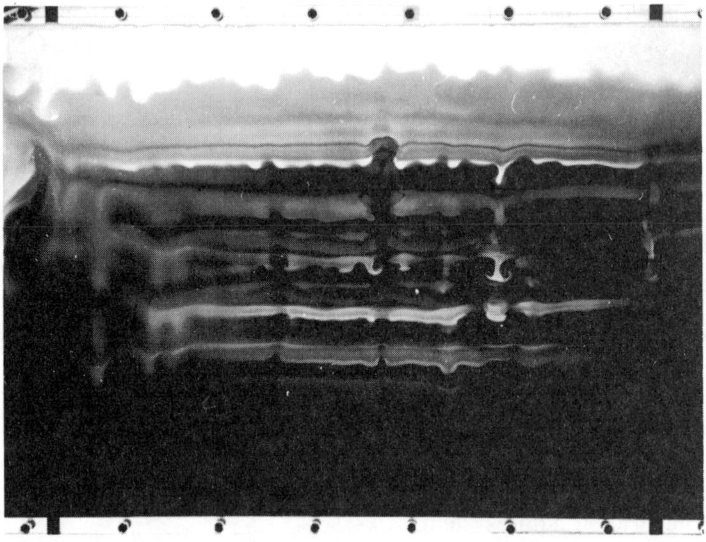

(a)

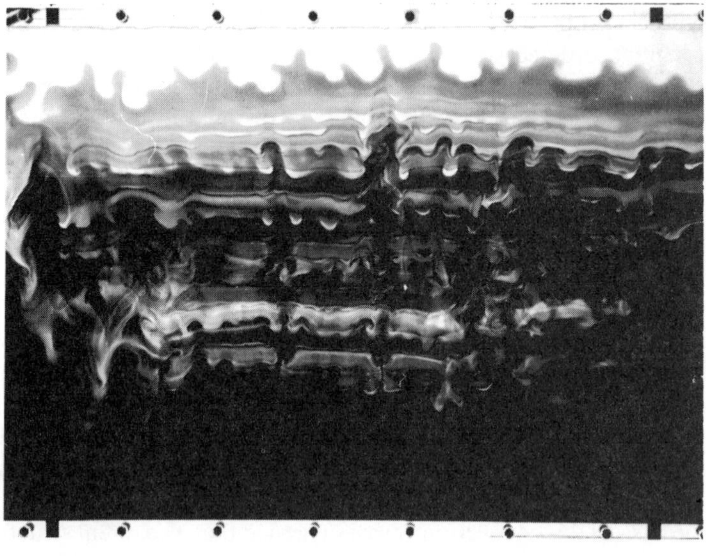

(b)

(c)
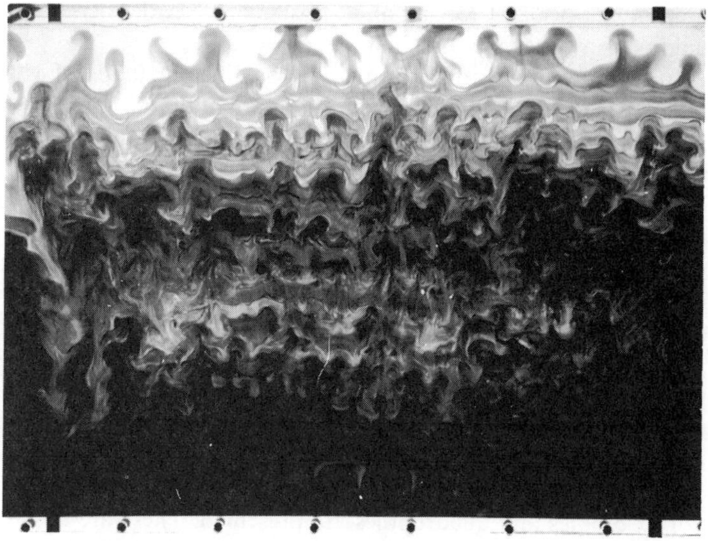

Figure 0.5 *Layering and subsequent formation of numerous dipoles in an overturned, initially stably stratified fluid. The initial buoyancy frequency* $\mathcal{N} = 3 \times 10^{-3}\,\mathrm{s}^{-1}$, *and the time after overturning* $t = 49\,\mathrm{s}$ *(a),* $55\,\mathrm{s}$ *(b) and* $62\,\mathrm{s}$ *(c).*

this general approach seems very promising, it requires the use of sophisticated mathematical techniques, and is outside the scope of this book.

In this book we consider the results of laboratory experiments concerning different vortex structures together with appropriate theoretical analyses. The book is organized as follows: To make clear what types of flows are considered, some simple experiments are described in Chapter 1 (section 1.1). Since the results of laboratory experiments are frequently used as the basis for theoretical analysis, it seems useful to give a short description of experimental techniques (section 1.2).

Chapter 2 contains a theoretical introduction. A reasonably comprehensive outline of the dynamics of vortex flows is presented, allowing the reader to understand the following chapters without the need to consult other textbooks. Particular attention is paid to vortex multipoles, localized forcing and self-similarity. A density-stratified fluid is also considered in Chapter 2.

In Chapter 3 a theoretical analysis of different kinds of vortex multipoles is presented. The general solution of the two-dimensional vorticity equation is considered. By means of a multipole expansion, the vorticity distributions obtained are connected with the appropriate systems of localized forces applied to a viscous fluid. This also helps in the introduction of the proper governing parameters, in particular the intensities of vortex multipoles, which are used for the interpretation and explanation of experimental results on vortex dipole interaction considered in Chapter 4.

Although the analytical solutions considered in Chapter 3 describe the mechanism of creation of vortex multipoles and qualitatively reproduce the subsequent flow evolution, these solutions fail to describe quantitatively the nonlinear dynamics of real vortex flows in a stratified fluid. Therefore in Chapter 5 empirical models for vortex dipoles and quadrupoles are presented. These are based on integral conservation equations and some physical hypotheses that can be justified by comparison with experimental results.

1

Introduction to experimental techniques

To illustrate the types of flows that are considered in this book, some simple experiments will be described in this chapter. These experiments are simple enough to perform in a household kitchen, without the need for any special equipment. It is useful to do this, since insight into a physical process is always improved if the dynamical evolution of a pattern produced by the process can be directly observed rather than just reconstructed mentally from a series of photographs.

In the following chapters photographs of flows as well as quantitative data obtained in laboratory experiments are frequently used in the consideration. In order for the reader to have a clear understanding of how these experimental data were obtained and also to stress their reliability, we also present here a short description of the experimental techniques. Although the general principles of laboratory experiments are rather simple, it is worth while to discuss some of them briefly.

1.1 Simple experiments

1.1.1 Decay of 'planar' turbulence in a stratified fluid

To observe the evolution of quasi-planar turbulent flow in a density-stratified fluid one first has to prepare a stratified fluid. For this experiment a shallow transparent dish, 20 cm × 30 cm × 5 cm, made from glass or Perspex, is also needed. In order to reproduce an almost-planar flow the dish is filled with two layers of fluid of different densities. The lower layer can be relatively thick, 2–3 cm, and consists of a heavy fluid: salt water with a concentration of salt 50–100 g per litre of water. The upper layer should be as thin as possible, 0.3–0.5 cm, and consists of a light fluid: fresh water. It is useful to add some drops of shampoo to the fresh water to reduce the surface tension and thus prevent the formation of a thin film on the surface of the upper layer. Such a film would play the role of an elastic cover and strongly oppose motion in the upper layer. To prepare a two-layer system without mixing the fluids, one can float a sheet of paper on top of the salt water. The fresh water is carefully poured onto the sheet, and the latter is then (very carefully) removed. The lower layer 'shields' the upper layer from the bottom, i.e. bottom friction is reduced by the presence of the lower layer.

Once the two-layer system has been prepared, the question arises as to how to excite the chaotic turbulent motion in the upper layer. To produce three-dimensional turbulent flow in a volume of fluid, a planar grid consisting of round or square rods is usually towed through it. To produce a two-dimensional flow, one can use a one-dimensional analogue of this grid: a comb made of thin wires, of diameter $d = 0.1$–0.2 cm and separation $d_0 = 0.5$–1.0 cm. This linear grid is placed vertically into the upper layer of fluid so that the ends of the wires do not touch the interface between the layers. The flow can be induced by towing the grid horizontally at a speed $U = 5$–$10\ \mathrm{cm\ s^{-1}}$ from one side of the dish to the other and then in the opposite direction. In this case the total momentum imparted to the fluid by the grid is zero and the resulting flow does not have a preferred direction.

Each wire of the moving grid produces a typical wake. If the Reynolds number $Re = Ud/\nu$ (where ν is the viscosity of the fluid: $\nu = 10^{-2}\ \mathrm{cm^2\ s^{-1}}$ for water) exceeds some critical value $Re_c \approx 50$, the wake is unstable and transforms into the well-known von Kármán vortex street. Numerous small vortices develop and interact with each other, forming in the upper layer a complex picture of irregular planar motion. Prior to the experiment one should add a few drops of ink or some other dye to the upper layer in order to visualize the flow pattern as in Figure 1.1(a). The picture has more contrast when the dish is illuminated uniformly from the bottom through a sheet of translucent paper.

On looking at the flow carefully, one can see that the moving grid generates not only intense horizontal motion in the upper layer but also some vertical motion at the interface between the two layers. As a result, internal waves appear in this region. However, these waves influence the flow only during the initial period of flow adjustment. As time progresses, the vertical motion rapidly decays, and the flow in the upper layer can be considered as nearly horizontal. This is the so-called intermediate asymptotic stage of flow evolution. During this stage vortices of opposite sign rapidly form vortex couples (Figure 1.1b). In contrast to single vortices, the couples, which are also termed vortex dipoles, possess linear momentum and can move. The moving vortex dipoles increase in size as a result of viscous entrainment of the ambient fluid, and they interact with each other through splitting of the vortices from the dipoles and subsequent merging of vortices of the same sign and pairing of vortices of opposite sign. As a result, the typical length scale of the dipoles increases with time, while the number of dipoles decreases (Figures 1.1c, d), and sometimes only one or two large dipoles survive in the flow.

The main aim of this simple experiment is to demonstrate how organized vortex structures appear in the irregular flow induced by localized forces acting in a viscous stratified fluid. The wires of the grid set the ambient fluid particles into motion from a state of rest. The fluid particles are thus accelerated, which means, according to Newton's second law of motion, that some external forces must be applied to the fluid. Because in a real viscous fluid there is a no-slip condition on the surfaces of the wires, the total external force applied to the fluid is equal in magnitude to the sum of the drag forces acting on the wires. These forces can be considered as localized ones, since the fluid volume on which they act is small compared with the distance between them. In this experiment the diameter of each wire is much less than the separation of the wires. The typical horizontal length scale of the vortex structures generated by the grid increases with time, and rapidly becomes large compared with the size of the fluid volume to which the forces were initially applied. It becomes clear from further analysis that asymptotically the structure of a vortex dipole does not depend strongly on the details of the initial force distribution in a local volume of fluid or on the particular initial conditions. Taking these facts into account, one can consider a vortex dipole to be a fundamental self-similar structure, appearing as the result of any localized forcing in a viscous stratified fluid. To

(a)

(b)

Figure 1.1 *Photographic sequence showing a top view of the evolution of a planar irregular vortex flow in the upper layer of a stratified fluid: (a) t = 2 s; (b) 8 s; (c) 36 s; (d) 96 s. The initial flow (t = 0) is induced by towing a grid of vertical rods in the upper layer. The row of points to the left of the first*

(continued on next page)

(c)

(d)

Figure 1.1 (*Contd.*)
photograph represents the horizontal cross-section of the grid, and the arrows show the directions of motion of the grid. The length scale of the motion (the typical diameter of the dipoles) is approximately 10 times greater than the depth of the upper layer, and increases with time as $t^{1/2}$.

demonstrate this more clearly, we shall consider another simple experiment.

1.1.2 Self-organization of a moving turbulent cloud

For this experiment one can use the same arrangement as in the first and try to create a large vortex dipole in a thin upper layer. The simplest way to produce a dipole is to use a short blast of air from a thin tube along the surface of the dyed upper layer (Figure 1.2). However, using this method, it is impossible to control the external force applied in the fluid.

A much better result could be obtained in a fluid with a continuous vertical distribution of density. A linearly stratified fluid, i.e. one where the density increases linearly with depth, is often used in experiments. The linear density distribution is created by linear variation of the salt concentration in the vertical direction. A simple arrangement called a two-tank system is usually used for this purpose. This arrangement is described in section 1.2.2.

Figure 1.2 *Vortex dipole induced in the upper layer of a stratified fluid by the impulsive action of a narrow air jet along the surface of the fluid.*

Suppose one has at one's disposal a large tank filled with a linearly stratified fluid. Now the problem is to reproduce a controllable action of a localized force in the fluid. Let us fix a thin horizontal nozzle (which can be a round glass or metal tube of diameter $d = 0.1$–0.2 cm) at some level inside the fluid and connect the nozzle by a flexible tube with a small burette. A small amount of fluid is then sucked through the nozzle into the burette so that the latter is filled with fluid of the same density as the fluid in a tank at the level of the nozzle. In order to visualize the flow, one can add a few drops of ink to the burette. By raising the burette slightly above the level of the water in the tank, one can inject the dyed fluid from the nozzle back into the tank. The injected fluid moves horizontally at the equilibrium density level and transports momentum into the fluid in the tank. By measuring the mass flux from the nozzle, one can accurately estimate the momentum flux produced by this source. Hence the intensity of the force acting locally upon the fluid is a controllable parameter. This is the so-called submerged jet technique.

To generate a strong dipole, a strong jet is needed. Such a jet is usually turbulent, and to produce it a small amount of fluid (1–2 cm^3) must be injected during a short period of time (2–4 s) from the nozzle. The injected conical jet becomes mixed with the surrounding fluid and rapidly transforms into a moving chaotic turbulent cloud (Figures 1.3a, b). Because of entrainment of the ambient fluid, the cloud of moving dyed fluid becomes much larger than the initial volume of injected fluid. Thus the small amount of injected fluid does not play a significant role in the flow dynamics. The main effect of injection is the transport of momentum into the surrounding fluid. As time progresses, the buoyancy force in a stratified fluid suppresses vertical motion in the moving cloud, and the flow becomes horizontal. Small-scale chaotic motions decay (Figures 1.3c, d). As a result, the cloud transforms into a large coherent vortex structure – a vortex dipole (Figure 1.3e).

We have already observed numerous dipoles in the previous experiment, but in that case they were not so well formed and symmetric as in the present experiment. Here a dipole moves forward, entraining ambient transparent fluid. This process leads to the formation of a typical spiral structure of vortices in the dipole. The horizontal dimension of the dipole increases. In contrast, the dipole's velocity eventually decreases with time, because momentum goes

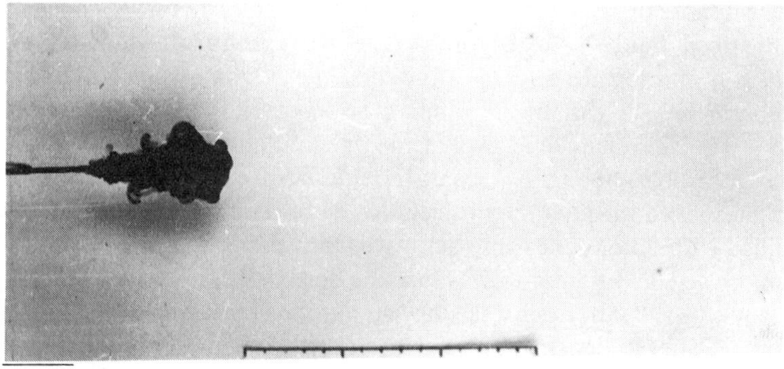

(a)

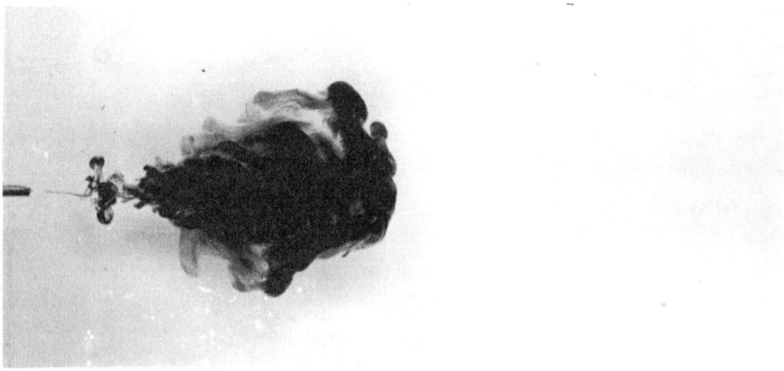

(b)

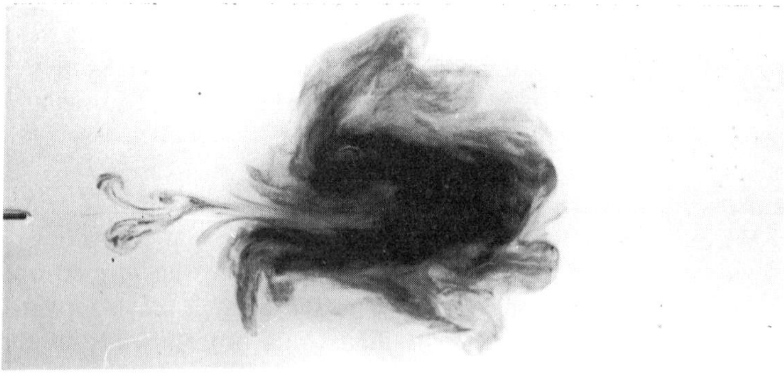

(c)

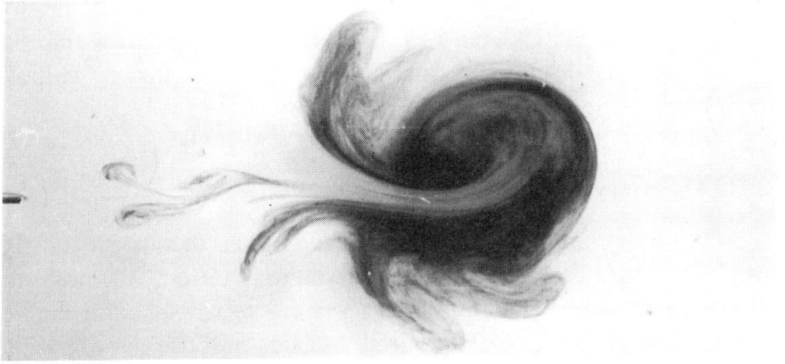

(d)

(e)

Figure 1.3 *Formation of a vortex dipole in a linearly stratified fluid. Initially* ($t = 0$) *a strong jet of dyed fluid is injected horizontally from a small round nozzle (diameter* $d = 0.2\,\text{cm}$*) during a short period of time* ($\Delta t = 5\,\text{s}$). *The irregular small-scale motions decay and a large coherent vortex structure forms: (a)* $t = 3\,\text{s}$*; (b) 5.5 s; (c) 12 s; (d) 22 s; (e) 57 s. The buoyancy frequency of a linearly stratified fluid is* $\mathcal{N} = 1.5\,\text{s}^{-1}$*; the initial Reynolds number of the flow* $Re = 4q_*/\pi\nu d = 720$*, scale in centimetres.*

into the acceleration of the initially quiescent and then entrained ambient fluid. The essential feature of the evolution of the dipole is that, after it is formed, its shape and internal structure (at least in the first approximation) remain similar at any instant of time. The idea that the evolution of vortex dipoles and other basic vortex structures can be considered as a self-similar process essentially simplifies the theoretical analysis. This idea will be broadly explored in the following discussions, and, where possible, checked experimentally.

1.2 Laboratory methods

The simple experiments considered in the previous section are mostly qualitative ones for the purposes of demonstration. To perform quantitative measurements of flow characteristics, controllable experimental conditions are required. For this purpose some specific experimental techniques have been developed. A few of these that are widely used in laboratory experiments on stratified fluids are discussed briefly below.

1.2.1 Experimental apparatus

A sketch of a typical experimental installation is shown in Figure 1.4. The experimental tank is usually made from glass or Perspex giving

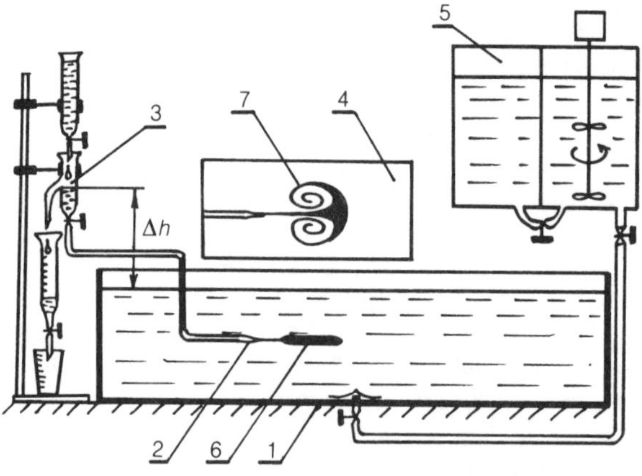

Figure 1.4 *Sketch of experimental apparatus used to study vortex structures in a stratified fluid: 1 = experimental tank with a stratified fluid; 2 = horizontal nozzle; 3 = water supply system with a constant-level bottle, Δh = const; 4 = 45° mirror; 5 = two-tank system (left tank, salt water, right tank, fresh water); 6 = side view of the flow; 7 = top view of the flow.*

transparent and optically homogeneous walls and bottom, which is important from the point of view of photographic or video recording of the flow and for observation of the flow by means of the shadowgraph technique. A spray head in the bottom of the tank serves for filling the latter with a fluid of inhomogeneous density while avoiding mixing.

Since the characteristics of the flow under investigation must not depend on the size of the tank, the latter must be sufficiently large. On the other hand, it should not be so big that the process of filling it with a large amount of stratified fluid becomes difficult. To find the optimal size of the tank, it is sometimes useful to make preliminary tests. First one of the main characteristics of the flow (e.g. its typical length scale) should be measured in a rather large tank. Then some additional vertical walls should be introduced into the tank and the measurements repeated. By decreasing the size of the tank in this way, one finds a lower limit at which the influence of the walls on the flow characteristics becomes comparable to the accuracy of measurements. Then the optimal size of the tank is taken as two to three times greater that its minimal size. Bearing in mind that to obtain reliable data a large number of experiments for different values of external parameters are usually needed, these preliminary tests generally pay for themselves with ease.

Most of the experiments discussed in this book were conducted in square (80 cm × 80 cm × 30 cm) or long (150 cm × 30 cm × 30 cm) tanks made from Perspex; in some experiments a shallow dish (40 cm × 30 cm × 5 cm) made from glass was used.

1.2.2 Creation and measurement of density stratification

A density-stratified fluid is a liquid system where vertical motion is suppressed by the gravitational force. There are also other liquid systems where some physical mechanism suppresses the motion in one direction and makes the flow quasi-two-dimensional. In a rotating homogeneous fluid the action of the Coriolis force leads to uniformity of the flow along the axis of rotation. One can reproduce a planar flow in an electrically conducting layer of mercury placed in a strong magnetic field. A soap film fixed on a wire frame has also been used to obtain a quasi-two-dimensional flow. However, all of these other methods are rather complicated, and have the essential shortcoming that in these systems it is very difficult to control accurately the value of the main governing parameter – the intensity of forcing. In contrast, this is not a problem in a stratified fluid (section 1.2.3).

The simplest case of a stratification is a two-layer system. In section 1.1 layers of salt and fresh water were used to make a stratification. This system cannot exist for long. With time, a density step is smoothed by diffusion of salt. Besides this, the layers eventually mix during the experiments. Instead of salt and fresh water, one can use immiscible liquids to create a two-layer stratification. A suitable pair of liquids is the dense ($\rho = 1.5\,\mathrm{g\,cm^{-3}}$) carbon tetrachloride ($CCl_4$) with low viscosity ($\nu_* = 5 \times 10^{-3}\,\mathrm{cm^2\,s^{-1}}$) and lighter but more viscous aqueous glycerol. The viscosity of an aqueous solution of glycerol depends strongly on the concentration of glycerol and can easily be varied over a wide range. A suitable value for our experiments is $\nu = (3–5) \times 10^{-2}\,\mathrm{cm^2\,s^{-1}}$, corresponding to a dilute solution. A two-layer system consisting of a thin (0.3–0.5 cm) layer of the prepared solution lying on a thick (2–3 cm) layer of carbon tetrachloride allows one to generate a nearly two-dimensional flow in the upper layer. Since the viscosity of the upper fluid is greater than that of the lower fluid, one can expect that a flow induced in the upper layer will be practically uniform throughout the depth of this layer. This means that the vertical flux of a horizontal component of momentum from the upper to the lower layer is negligible compared with the transport of momentum by the flow in a horizontal direction in the upper layer. Thus the flow in the upper layer appears to be 'shielded' from the bottom of the tank, i.e. the bottom friction is effectively reduced by the presence of the lower layer. As mentioned previously, a few drops of shampoo added to the upper fluid reduces the effects of surface tension on the interface between the layers and on the surface of the fluid.

Another type of stratified system is a continuously stratified fluid. This kind, in particular one with a linear vertical density distribution, is widely used in experiments where quasi-two-dimensional flows are studied. Theoretical analysis of the governing equations (section 2.3.3) shows that a whole class of flows induced by the action of localized forces in a linear stratified fluid has a typical intermediate asymptotic regime in which vertical motion is strongly suppressed by the buoyancy force and the flow becomes horizontal. Fluid particles in these flows move in horizontal planes, and the ambient fluid is entrained mostly horizontally into the flow. Neglecting the effects of vertical entrainment, the flow can be considered as quasi-two-dimensional and localized in some horizontal layer in the bulk of the stratified fluid. The important advantage of this system compared with that consisting of two immiscible fluid layers is the absence of surface tension effects.

To create a linearly stratified fluid, a simple two-tank arrangement is used. This is a tank divided by a vertical wall into two equal parts, connected to each other by a flexible tube with a tap (Figure 1.4). The right-hand tank contains fresh distilled water, while the left-hand one contains salt solution of the maximum density required (10–100 g of salt (NaCl) per litre of distilled water). The right-hand tank is connected by a flexible tube with a spray head at the bottom of the experimental tank. A mixer with two or three propellers is installed in the right-hand tank. To fill the experimental tank with linearly stratified salt water, one turns on both taps and switches on the mixer. Initially, fresh water from the right-hand tank begins to flow into the experimental tank and spreads uniformly on the bottom. At the same time, salt water from the left-hand tank begins to run into the right-hand one because of the hydrostatic pressure difference, and then mixes there with fresh water. Thus the water in the right-hand tank becomes saltier with time. This salt water flows into the experimental tank and displaces the less salty water, creating a stratification. It is easy to show that if the water flux from one tank to another is constant with time and is equal to one-half of the flux from the two-tank system to the experimental tank then the salinity of the water in the right-hand tank increases linearly with time; hence the salinity of the water in the experimental tank increases linearly with depth. This two-tank system allows the creation of both strong and weak stratifications. The weaker the stratification, the more carefully should one fill the tank. A piece of foam rubber inserted into the spray head is very useful for reducing mixing during the filling of the tank.

In principle, by changing the fluxes through both taps and using a simple theory proposed by Oster (1965), one can create any desirable stable distribution of salt concentration in water with depth.

By definition, the salinity s of water is the non-dimensional quantity

$$s = \frac{m_s}{m_s + m_w}, \tag{1.2.1}$$

where m_s is the mass of salt and m_w the mass of pure water in the solution. Usually the salinity S is measured in promille (parts per thousand, ‰), so that $S = 1000s$. Over a wide range of salinity values (5–100‰), with good accuracy the density ρ of salt water depends linearly on salinity when the temperature T remains constant:

$$\rho = \rho_0(1 + \beta S), \tag{1.2.2}$$

where $\beta = 8 \times 10^{-4}$ (‰)$^{-1}$ and $\rho_0 = 1\,\mathrm{g\,cm^{-3}}$ for $T = 20°\mathrm{C}$. Nowadays it is unusual to measure the salinity of sea water directly with high accuracy. For the purposes of laboratory experiments it is more convenient to measure the conductivity σ of the solution and then to calculate the salinity using the simple empirical relation (e.g. Martin, Simmons and Wunsch, 1972)

$$\sigma = \sigma_0 + \gamma S, \tag{1.2.3}$$

where $\sigma_0 = 2 \times 10^{-3}\,\mathrm{cm^{-1}\,\Omega^{-1}}$ and $\gamma = 1.5 \times 10^{-3}\,\mathrm{cm^{-1}\,\Omega^{-1}}$ (‰)$^{-1}$.

For conductivity measurements different types of probes are used. The simplest is a single-electrode probe (Gibson and Schwarz, 1963; Afanasyev, Voropayev and Filippov, 1990), which is a small platinum sphere (diameter $d = 10^{-2}$ cm) covered with platinum black and fixed at the end of a thin glass capillary. The spatial resolution of the probe, i.e. the radius of some effective fluid volume that gives for example 90% of the measured value of conductivity, is approximately equal to $5d = 5 \times 10^{-2}$ cm. When made carefully, the probe is stable and gives a low noise level. It allows values of density to be calculated from (1.2.2) and (1.2.3) with an accuracy $\Delta\rho/\rho \approx 10^{-5}$ that is usually sufficient for the purposes of laboratory experiments. The probe can be used for traversing the fluid volume, giving a density distribution along its path, and also for measuring density pulsations at a fixed point of space. Typical examples of such records are shown in Figures 5.5 and 5.8.

1.2.3 Localized sources

Most of the flows discussed in this book are induced by a localized forcing of fluid from a state of rest. Moreover, each flow in its archetypal form is induced by a particular localized source. This may be a single force, a force dipole where two equal forces act in opposite directions, a sink or source of mass, a source of angular momentum, or a combination of such sources. To simplify the following theoretical analysis, all sources will be considered to be localized in space, i.e. the size of the volume on which the source acts is small compared with the typical length scale of the resulting flow. Thus the question arises as to how to reproduce different kinds of controllable forcing experimentally.

In the experiment described briefly in section 1.1.2 a submerged jet was used to reproduce the action of a single force. Let us consider this method in more detail. A jet from a thin round nozzle (diameter d) is injected into the surrounding fluid. If the mean velocity at the nozzle exit is U_* then the mass flux ρq_* is proportional to $q_* \approx \frac{1}{4}\pi d^2 U_*$. The injected fluid transports momentum, the momentum flux ρJ being proportional to

$$J \approx \tfrac{1}{4}\pi d^2 U_*^2 = \frac{4q_*^2}{\pi d^2}. \tag{1.2.4}$$

According to Newton's second law of motion, the rate of change of the momentum of the system is equal to the magnitude of the force applied. Thus the source acts on the fluid with a total force ρF, with $F = J$, and one can estimate the value of the force acting on the fluid from the measured value of the mass flux ρq_*.

Suppose that we decrease the diameter d of the nozzle and increase the velocity U_* of injection such that the product dU_* remains constant. In this case q_* tends to zero while J remains constant. Thus we obtain asymptotically a 'point' source of momentum. To model the action of a point force on the fluid, one must use a nozzle with a small diameter ($d = 0.05$–0.2 cm). Hence the mass flux through the nozzle must also be small. A simple water supply system can be used to produce a controllable flux. This system (Figure 1.4) consists of two glass burettes with 0.05 ml scales and a constant-level bottle, which is fixed on a vertically movable support. By lowering and raising the bottle, one can vary the difference Δh between the water level in the tank and that in the constant-level bottle, and thereby very q_*. This system provides a small flux of water that is very stable in time. By measuring the total volume of fluid before and after the experiment, one can measure the mass flux ρq_* with good accuracy, and hence calculate the momentum flux ρJ using (1.2.4).

The considered point source of momentum is an axisymmetric one. Using the same technique, one can reproduce a line source of momentum to induce a planar flow in a liquid layer. A thin slitted nozzle should be used for this purpose. The height of the nozzle has to be equal to the depth h of the fluid layer ($h = 0.3$–0.5 cm). The typical width d of the slot is 0.01–0.03 cm. A momentum flux ρJ per unit height for a line source of momentum can be estimated using a dependence similar to (1.2.4):

$$J = q_*^2/dh^2. \tag{1.2.5}$$

Note that in (1.2.4) and (1.2.5) a uniform velocity distribution at the nozzle exit is used. We do not know the real velocity distribution. However, it is easy to show that if the distribution is not uniform then an insignificant constant of order unity will appear in (1.2.4) and (1.2.5). For example, a parabolic distribution gives a constant equal to $\frac{4}{3}$ in (1.2.4).

To reproduce more complicated forcing on the fluid, one can use two or more submerged jets. For example, the action of two equal forces of opposite direction (a force dipole) can be modelled by two jets from two nozzles directed towards one another. A typical quadrupolar flow produced by this source of motion in a thin layer of fluid is shown in Figure 1.5. It will be shown below that the main governing parameter for this type of source is the intensity ρQ of the line force dipole, defined by

$$Q = \varepsilon J,$$

where ε is the distance between the nozzles.

The quadrupolar flow presented in Figure 1.5 was produced when the nozzles were placed in juxtaposition. If the nozzles are fixed in a solid supporting frame that permits their precise positioning and orientation, it is easy to demonstrate that when the nozzle axes are offset by a small distance δ, the two induced dipoles move away at some angle to the direction of the nozzles. This latter feature is illustrated in the sequence of photographs in Figure 1.6. When $\delta \neq 0$, an additional governing parameter $M = \delta J$ appears. This parameter characterizes the intensity of the force pair that supplies an angular momentum ρM to the fluid and causes rotation of the flow.

In the course of performing a number of experiments, it was felt, however, that this generation technique has a serious drawback. The combination of two nozzles is not really a localized 'point' source of motion, because the distance ε between the nozzles is not small enough compared with the typical length scale of the resulting flow. If one tries to reduce ε significantly, the resulting flow becomes very unstable. Besides, the nozzles have finite dimensions and are thus likely to influence flow. In spite of these shortcomings, the method of two jets is very useful in qualitative experiments for the purpose of demonstrations. However, to obtain quantitative information, it is better to use other methods of flow generation. A method that is free from the above-mentioned shortcomings is based on the fact that any solid body moving in a real viscous fluid acts on the fluid with a force equal to the drag force on the body.

When a solid body oscillates back and forth about some mean position in a fluid, it exerts a force on the latter in alternating directions. It can be shown that the resulting force varies periodically with the frequency of oscillation of the body and consists of two parts: an inertial one and a dissipative one. The phase shift between the inertial force and the velocity of the body is $\frac{1}{2}\pi$, and therefore the inertial force does not increase the kinetic energy of the fluid. In contrast, the dissipative force that appears owing to the no-slip conditions at the body surface and the subsequent creation of a boundary layer on the body varies with the same phase as the velocity of the body.

In some simple cases it is possible to estimate accurately the amplitude of the dissipative force. For example, for a cylinder of diameter d oscillating with a small amplitude ε in a direction perpendicular to its axis the mean amplitude of this force (per unit length of the cylinder) is given by (Batchelor, 1967)

$$\rho J = 4\pi^{5/2}\rho d \nu^{1/2}\varepsilon f^{3/2},$$

where f is the frequency of oscillation (measured in hertz), ρ is the density and ν is the viscosity of the fluid. For a symmetric body the mean value of the force over the period of oscillations is zero. Thus one can try to use a small oscillating cylinder as a localized source of motion to reproduce controllably the action of a line force dipole. Indeed, by application of this technique, it is possible to reproduce quadrupolar flow structures as presented in Figure 1.7.

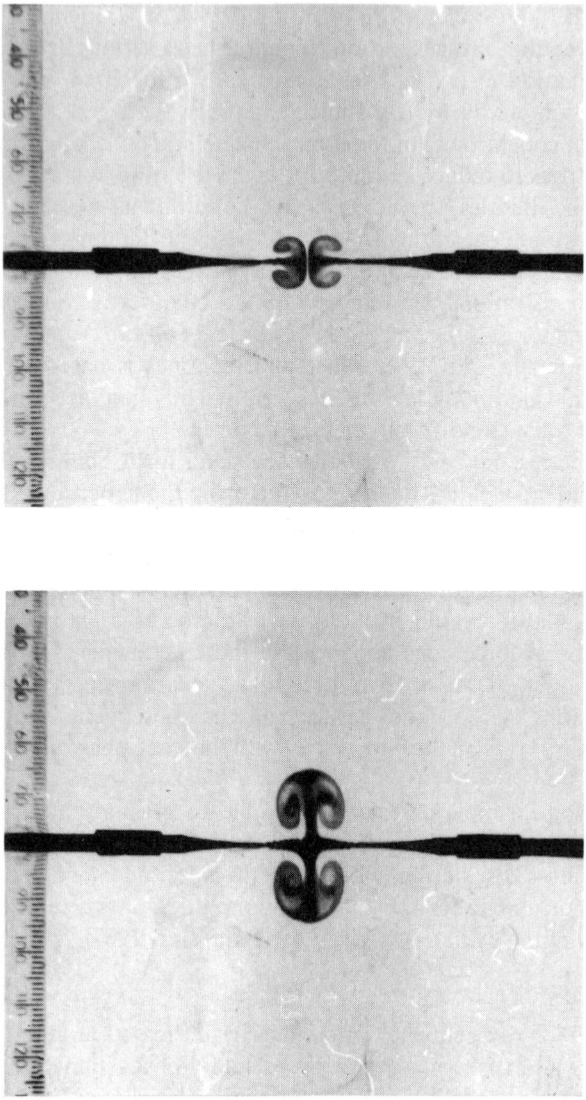

(a)

(b)

Figure 1.5 *Sequence of plan-view photographs showing the formation and subsequent evolution of a symmetric vortex quadrupole in a thin upper fluid layer. The flow is driven by two jets from slitted nozzles directed towards one another. The flux rate of the dyed fluid from each nozzle is* $q_* = 6 \times 10^{-2}\ cm^3\ s^{-1}$, *the*

(c)
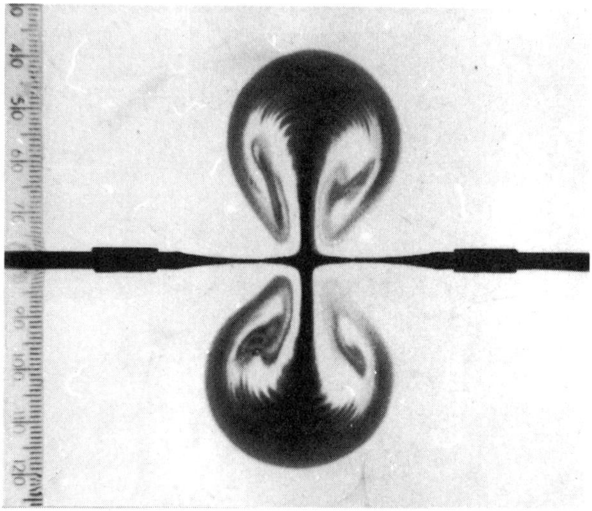

cross-sectional area of each nozzle is $h \times d = 0.4\,\text{cm} \times 0.03\,\text{cm}$*, and the distance between the nozzles is* $\varepsilon = 1.75\,\text{cm}$*. The numbers on the scale represent millimetres. The photographs were taken at times* $t = 0.5\,\text{s}$ *(a),* $3\,\text{s}$ *(b), and* $30\,\text{s}$ *(c) after the forcing started.*

The arrangement of the generation technique consists of a rigid horizontal frame connected to a loudspeaker, whose motion can be controlled by a function generator. The cylinder (a thin steel rod of diameter $d = 0.03$–$0.16\,\text{cm}$) is fixed to this supporting frame and is placed vertically in the upper fluid layer (Figure 1.7) such that its end almost touches the interface between the layers. In this way it is possible to allow the cylinder to perform horizontal oscillations in a sinusoidal fashion with adjustable frequency and amplitude.

Thus the action of a small oscillating cylinder is similar to the action of a point force dipole. For quantitative characterization of the source intensity in experiments the following estimate can be used:

$$Q = \varepsilon J = 4\pi^{5/2}\, d\nu^{1/2}\varepsilon^2 f^{3/2}. \tag{1.2.6}$$

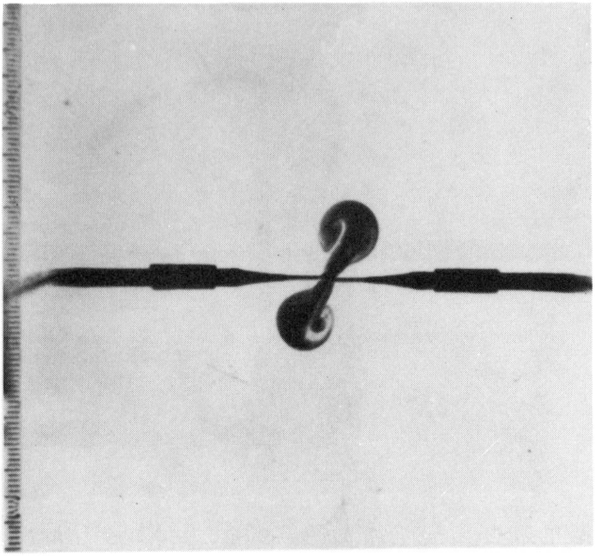

(a)

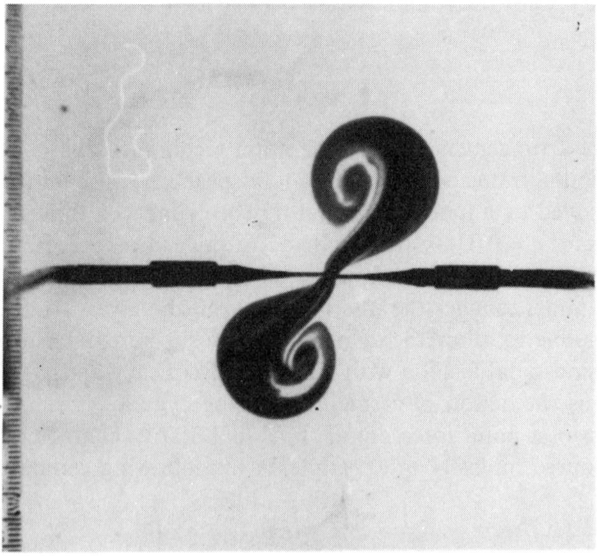

(b)

(c)

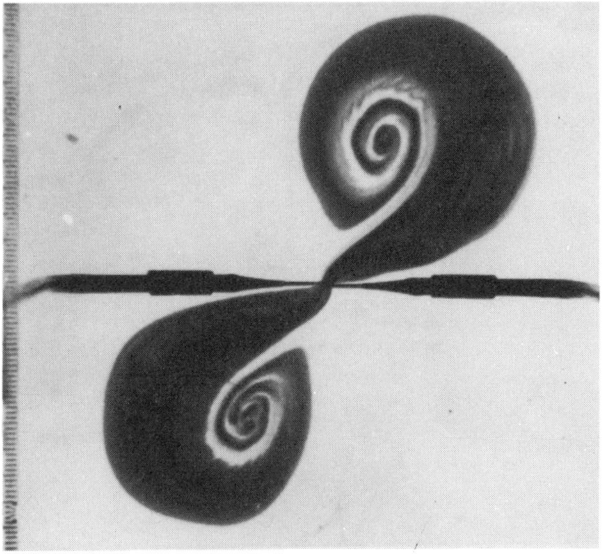

Figure 1.6 *Sequence of photographs showing the formation and subsequent evolution of a vortex quadrupole when the nozzle axes were offset by a small distance* $\delta = 0.05$ *cm: (a)* $t = 0.5$ *s; (b) 4 s; (c) 13.5 s. The experimental parameters are* $q_* = 6 \times 10^{-2}\ cm^3\ s^{-1}$ *and* $\varepsilon = 0.3$ *cm.*

In general, by placing the oscillating cylinder on light bearings and rotating it with angular velosity Ω using a small electric motor via a flexible connection, it is possible to transport to the fluid, in addition to Q, some angular momentum $\rho\mathcal{M}$ per unit time given by (see section 3.2.1)

$$\mathcal{M} = \pi \nu d^2 \Omega. \quad (1.2.7)$$

Note that the dimensions of Q and $\mathcal{M}$ are the same.

The oscillating body technique permits the reproduction of more complicated force actions on a fluid. For example, a towed grid as used in the experiment described in section 1.1.1 can be fixed to a supporting oscillating frame and placed vertically in the upper fluid layer. By oscillating the grid in horizontal directions in the same way as the cylinder, it is possible to generate a line of quadrupolar structures, thus modelling the interactions of vortex structures and

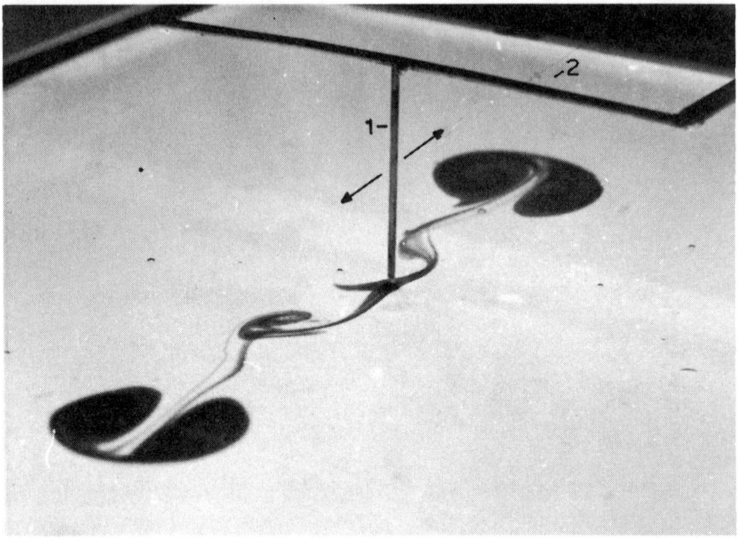

Figure 1.7 *Oblique view of the planar flow (visualized by dark dye) induced by a horizontally oscillating vertical cylinder (1), that was oscillated impulsively during a short period of time ($\Delta t = 4$ s) in the directions indicated by the arrows. The cylinder is fixed to a supporting frame (2) connected to an excitator. The frequency and the amplitude of oscillations are $f = 18$ Hz and $\varepsilon = 0.1$ cm respectively. The diameter of the cylinder is $d = 0.16$ cm, and the time is $t = 14$ s after forcing started.*

the propagation dynamics of the resulting quasi-two-dimensional chaotic flow.

Note, finally, that the localized sources of motion considered here can act in two regimes: impulsively during a short period of time, and with constant intensity during a prolonged period of time. The various forms of flows induced by the various forcings are considered in detail in the following sections.

1.2.4 *Flow visualization and velocity measurements*

Most fluids are transparent homogeneous media, and therefore to make their motion visible one must introduce some tracers into the flow. By observing the motion of tracers, one can obtain an idea of the flow development. The injection of dye has long been a popular method of visualizing flow patterns. One can use ink, food colouring or other kind of dye to mark fluid elements. However, this method has some shortcomings. The dye eventually contaminates the working fluid, and the tank has to be emptied and refilled after each experiment. Besides, the density of the dye solution usually differs from that of the working fluid.

A more appropriate technique for flow visualization employs the pH indicator thymol blue. This is orange-yellow in an acidic environment but changes colour to deep blue when the solution becomes alkaline. The working fluid is prepared by dissolving 0.1–0.2 g thymol blue per litre of distilled water and adding a few drops of hydrochloric acid. A homogeneous orange-yellow solution is obtained. Using this solution, a density stratification is created, employing the standard two-tank method. One or two drops of sodium hydroxide are added to the volume of fluid whose flow is to be visualized, so that it becomes deep blue. The flow is clearly visible, and since the density of the dyed fluid remains practically unchanged the buoyancy effects are negligible. After the end of the test one has to wait several minutes until the injected blue fluid in the tank becomes yellow again by a diffusive chemical reaction with the acidic surroundings. Thus it is possible to perform the next experiment without changing the working fluid in the tank. This method turns out to be very convenient, and most of the photographs presented in this book were produced using thymol blue.

The thymol blue visualization technique can also be exploited to measure the velocity distribution in a flow (Baker, 1966). This technique is applicable for low-speed flows with a range of velocities 0–5 $cm\,s^{-1}$. Two electrodes are introduced into the fluid. The negative electrode is a fine platinum wire (diameter 10^{-3} cm) or a cross of wires placed in the region of the flow to be measured. Any metal part of the tank or a copper plate can be used as the second, positive, electrode. When a DC voltage is applied between the electrodes, positive hydrogen ions migrate to the negative electrode, where they give up their charge and combine to form H_2 molecules. The solution therefore exhibits an excess of OH^- ions near the cathode and becomes basic. In a slightly acidic solution close to the point of becoming basic the shift in pH near the negative electrode towards the basic side causes the solution to change colour to blue.

(a)

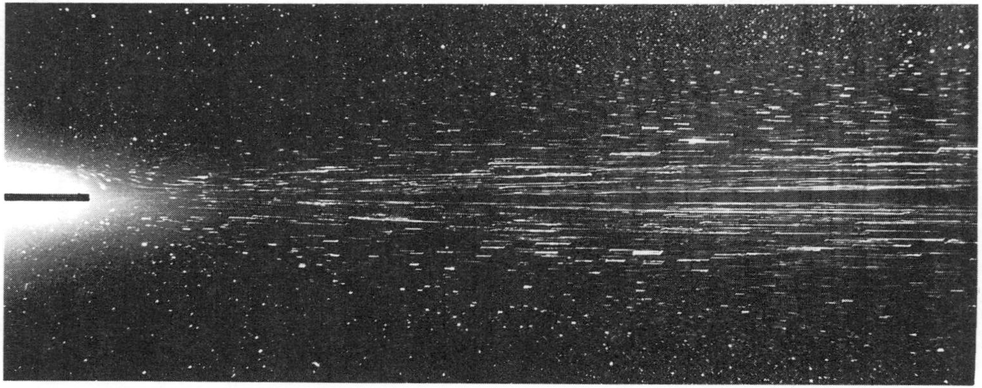
(b)

If the electric voltage is pulsed, small cylinders of dark blue fluid are formed around the wires and are swept along by the flow. By repeating the electric pulses, one obtains on a photograph well-defined dye lines (Figure 5.4b). From the position of these dye lines, and with a known pulse frequency, one can derive the velocity distribution in the flow under investigation. Note that when this method is used for a stratified fluid, the density stratification must not be produced by salt, which significantly increases the conductivity of the solution, leading to generation of bubbles at the electrodes through electrolysis. Instead of salt, sugar can be dissolved in distilled water to produce a stratified fluid. The typical current between the electrodes should not be much more than 10–15 mA when the applied voltage is about 10 V.

In some cases small foreign particles are used for flow visualization and velocity measurements. The particles have to be small enough to move along with the flow. In general, the particles do not follow the flow, but in some cases – in particular for steady flow – the difference between particle and fluid velocities is insignificant. The most frequently used particles are aluminium powder and small polystyrene spheres of neutral density. In the flow under study a thin plane slice is isolated for observations and measurements by means of a strong beam of light from a slit. For this purpose a laser beam split by a cylindrical lens into a thin sheet of light can be used. Particles moving in the illuminated plane scatter the light and give sharply terminated streaks on photographs (Figure 1.8). The lengths of tracks divided by the exposure time give the distribution of velocity in the illuminated plane.

Consider finally a simple shadowgraph technique that permits observation of density inhomogeneities in transparent media. A

Figure 1.8 *Streakline photograph of a steady submerged jet in a linearly stratified fluid visualized with aluminium powder: (a) top view; (b) side view. Experimental parameters are* $\mathcal{N} = 0.7\,s^{-1}$ *and* $Re = 145$*, and the exposure is* 6 s. *The black horizontal line on the left is the nozzle. The slow flow towards the source at large distances from the jet axis is a compensatory flux due to the finite dimensions of the experimental tank; the velocity distributions (Figure 2.3) were measured in a narrow zone near the axis of the jet where the influence of the backward compensatory flow is negligible.*

simple set-up with a light source and a lens (to make the light beam parallel) is shown in Figure 1.9. A parallel light beam goes through the test section of the experimental tank, producing a shadow pattern on the screen behind the tank. This method is based on the following physical principle. In general, the refractive index n of a transparent fluid changes with the fluid density. Thus a flow in a density-stratified fluid creates an inhomogeneous refractive field. On passing through such an inhomogeneous medium, a light ray is refracted and deflected from its original direction. According to Fermat's principle, the variation of optical path along a light ray must vanish. On this basis, it can be shown that the curvature of the optical path is proportional to the first derivative of the refractive index $\partial n/\partial z$ across the path. Thus if $\partial n/\partial z$ is constant along the test section then the deflection angle of all rays is the same; the rays remain parallel and the screen is uniformly illuminated. This is the case when a linearly stratified fluid, where $\partial n/\partial z = \text{const.}$, is not disturbed. When the fluid is disturbed, the derivative $\partial n/\partial z$ is not constant locally, i.e. the second derivative $\partial^2 n/\partial z^2$ is non-zero; hence rays going through these inhomogeneities are deflected more or less than those going through regions where $\partial n/\partial z$ is constant. As a result, the disturbances produce light and dark patterns on the screen (Figure 1.9). The image sharpness on the screen is proportional to f/dl, where f is the focal length of the lens, d is the diameter of the light source and l is the distance between the screen and the test section.

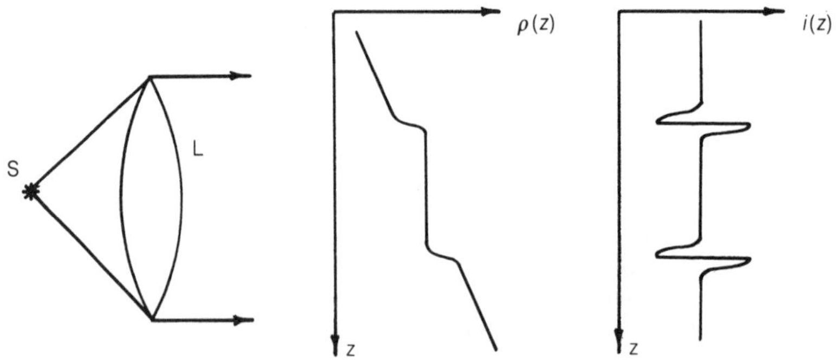

Figure 1.9 *Schematic arrangement of a simple shadowgraph system: S = light source; L = lens; $\rho(z)$ = vertical density profile in the tank; $i(z)$ = light intensity on the screen.*

With this simple technique, it is not possible to obtain quantitative data on the density field; nevertheless it does yield useful qualitative information on the flow field in addition to other methods of visualization (Figure 5.4a).

In this chapter we have described briefly some general laboratory methods – in particular, those that can be used to reproduce and to study under controllable conditions basic vortex structures in a stratified fluid. We have also tried to give a qualitative description of the flows under study by illustrative examples to clarify the main subject of this book.

Those interested in more information on methods of flow visualization and measurements should refer to the book by Merzkirch (1974).

2

Introduction to vortex dynamics

2.1 Equations of motion and elements of vortex dynamics

The main quantities determining the state of a moving fluid are the velocity vector $\boldsymbol{u} = (u_1, u_2, u_3)$ and two thermodynamic quantities, for example the pressure p and the density ρ. It can be shown that all the thermodynamic quantities can be determined from the equation of state when any two of them are known (Landau and Lifshitz, 1980). Thus the three velocity components, the pressure and the density comprise a complete set of quantities for a moving fluid. In general all of these are functions of the position vector $\boldsymbol{x} = (x_1, x_2, x_3)$ of a point in space, and also of time t.

Considering the coordinates of a point in space and time as independent variables, one obtains immediately that the rate of change of some fluid property $\Phi(\boldsymbol{x}, t)$ at a fixed spatial point $\boldsymbol{x}$ is by definition simply the partial derivative $\partial\Phi/\partial t$.

2.1.1 *Material fluid elements*

Consider now some fluid element that moves when the fluid is in a state of motion. Since the ambient fluid particles can move away from each other, one must consider a very small fluid volume of fixed identity. This volume is called a material fluid element. If the position of this element at time t is $\boldsymbol{x} = \boldsymbol{x}(t)$ then the property Φ of the element can be expressed as

$$\Phi = \Phi(\boldsymbol{x}(t), t).$$

This gives the rate of change of Φ for a moving fluid element as

$$\frac{\mathrm{d}\Phi}{\mathrm{d}t} = \frac{\partial\Phi}{\partial t} + \frac{\partial\Phi}{\partial x_1}\frac{\mathrm{d}x_1}{\mathrm{d}t} + \frac{\partial\Phi}{\partial x_2}\frac{\mathrm{d}x_2}{\mathrm{d}t} + \frac{\partial\Phi}{\partial x_3}\frac{\mathrm{d}x_3}{\mathrm{d}t} \equiv \frac{\partial\Phi}{\partial t} + \frac{\partial\Phi}{\partial x_i}\frac{\mathrm{d}x_i}{\mathrm{d}t}. \quad (2.1.1)$$

Since the velocity of the material element is by definition the rate of change of the position of this element,

$$\boldsymbol{u} = \frac{\mathrm{d}\boldsymbol{x}}{\mathrm{d}t},$$

(2.1.1) becomes

$$\frac{\mathrm{d}\Phi}{\mathrm{d}t} = \frac{\partial \Phi}{\partial t} + (\boldsymbol{u} \cdot \boldsymbol{\nabla})\Phi \equiv \frac{\mathrm{D}\Phi}{\mathrm{D}t}, \tag{2.1.2}$$

where

$$\boldsymbol{\nabla} = \left(\frac{\partial}{\partial x_1}, \frac{\partial}{\partial x_2}, \frac{\partial}{\partial x_3}\right)$$

is the gradient operator, $(\boldsymbol{u} \cdot \boldsymbol{\nabla}) = u_i \boldsymbol{\nabla}_i$ is the scalar (or dot) product, and a sum over repeated index is implied.

The additional term $(\boldsymbol{u} \cdot \boldsymbol{\nabla})\Phi$ is usually called the advective derivative, because it is the change in Φ as a result of movement of the fluid element from one location to another. A simple example helps to clarify the meaning of this term. Consider steady flow in a pipe of variable cross-sectional area. In steady flow all properties related to fixed spatial points do not change in time. Hence at a fixed spatial point we have $\partial \boldsymbol{u}/\partial t = 0$. On the other hand, it is known that the velocity is different in different sections of the pipe. Thus the velocity of a moving fluid element varies with time, because

$$\frac{\mathrm{D}\boldsymbol{u}}{\mathrm{D}t} = (\boldsymbol{u} \cdot \boldsymbol{\nabla})\boldsymbol{u} \neq 0.$$

2.1.2 *Conservation of mass*

It is postulated that fluids obey conservation laws for some basic physical quantities, namely mass and momentum. Thus the governing equations that determine the behaviour of a moving liquid express in fact the balance of these quantities.

In deriving the governing equations, one can use two equivalent methods. The first relates to a volume fixed in space, while the second relates to a material volume that consists of the same fluid particles and whose bounding surface moves with the fluid. In the following a fixed volume will be denoted by V and a material volume by τ.

Consider a volume V fixed in space. One can imagine it as a region of space through which the fluid flows freely. For simplicity suppose that there are no sources or sinks of conserved quantities in the fluid. It is obvious that in this case a rate of change of some conserved quantity in the considered volume is equal to the total inflow of this quantity through the bounding surface S of the volume. Here we take the vector surface element $\mathrm{d}\boldsymbol{S}=(\mathrm{d}S_1, \mathrm{d}S_2, \mathrm{d}S_3)$, of magnitude $\mathrm{d}S$ equal to the area of the surface element, to be directed along the outward normal to the surface.

One of the important conserved quantities is mass. The mass of fluid bounded by the closed surface S fixed in space is an integral over the volume V:

$$\int_V \rho \, \mathrm{d}V,$$

where $\rho = \rho(\boldsymbol{x}, t)$ is the spatial distribution of density in the considered volume at time t. The surface integral

$$-\int_S \rho \boldsymbol{u} \cdot \mathrm{d}\boldsymbol{S}$$

represents the total flux, i.e. the rate of inflow through the boundaries. Therefore the conservation of mass can be expressed in the integral form

$$\frac{\mathrm{d}}{\mathrm{d}t}\int_V \rho \, \mathrm{d}V = -\int_S \rho \boldsymbol{u} \cdot \mathrm{d}\boldsymbol{S}. \tag{2.1.3}$$

Transforming the surface integral on the right-hand side to a volume integral by means of the divergence theorem,

$$\int_S \rho \boldsymbol{u} \cdot \mathrm{d}\boldsymbol{S} = \int_V \boldsymbol{\nabla} \cdot (\rho \boldsymbol{u}) \, \mathrm{d}V,$$

and taking the time derivative inside the integral sign on the left-hand side since the volume V is fixed, one can rewrite (2.1.3) in the form

$$\int_V \left[\frac{\partial}{\partial t}\rho + \boldsymbol{\nabla} \cdot (\rho \boldsymbol{u})\right] \mathrm{d}V = 0. \tag{2.1.4}$$

This holds for any volume, which is possible only if the integrand vanishes at every point. Thus

$$\frac{\partial \rho}{\partial t} + \nabla \cdot (\rho \boldsymbol{u}) = 0, \tag{2.1.5}$$

which is called the continuity equation because it implies that the fluid has no voids in it. Equation (2.1.5) expresses in differential form the principle of conservation of mass.

Using the notation (2.1.2) for the substantial derivative $\mathrm{D}/\mathrm{D}t$, (2.1.5) can be rewritten in the form

$$\frac{1}{\rho}\frac{D\rho}{Dt} + \nabla \cdot \boldsymbol{u} = 0. \tag{2.1.6}$$

To understand the physical meaning of the velocity divergence $\nabla \cdot \boldsymbol{u}$, consider a volume τ consisting of moving fluid particles. Since the surface elements of the volume move with the fluid, the rate of change of τ is the surface integral

$$\frac{\mathrm{d}\tau}{\mathrm{d}t} = \int_S \boldsymbol{u} \cdot \mathrm{d}\boldsymbol{S},$$

which can be transformed into a volume integral $\int_\tau \nabla \cdot \boldsymbol{u}\, \mathrm{d}\tau$. Considering the relative rate of change of τ, as $\tau \to 0$, one obtains

$$\lim_{\tau \to 0} \frac{1}{\tau}\frac{\mathrm{d}\tau}{\mathrm{d}t} = \lim_{\tau \to 0} \frac{1}{\tau} \int_\tau \nabla \cdot \boldsymbol{u}\, \mathrm{d}\tau = \nabla \cdot \boldsymbol{u}.$$

A fluid is called incompressible if its density does not change with pressure. Therefore in an incompressible fluid the density of the moving fluid element does not change along the path of this element. This implies that

$$\frac{\mathrm{D}\rho}{\mathrm{D}t} \equiv \frac{\partial \rho}{\partial t} + \boldsymbol{u} \cdot \nabla \rho = 0 \tag{2.1.7}$$

and the equation of conservation of mass (2.1.5) for an incompressible fluid becomes

$$\nabla \cdot \boldsymbol{u} = 0. \tag{2.1.8}$$

An incompressible fluid is called homogeneous in density if $\rho = \text{const}$. For such a homogeneous fluid (2.1.7) is satisfied identically, giving (2.1.8).

2.1.3 *Equations of motion*

In addition to mass, momentum is also postulated to be a conserved quantity for fluid media. The mass is a scalar quantity; hence the flux of mass is a vector, which can also be considered as a first-order tensor. Since the momentum is a vector quantity, the flux of momentum is a second-order tensor. It is useful to introduce the second-order tensor Π_{ik}, which represents the flux of the ith component of momentum through unit area of surface fixed in space and oriented normally to the direction of the coordinate axis x_k. In textbooks on fluid dynamics it is shown that to satisfy the law of conservation of angular momentum Π_{ik} must be a symmetric tensor: $\Pi_{ik}=\Pi_{ki}$. For an ordinary viscous fluid the tensor Π_{ik} can be expressed as (Landau and Lifshitz, 1987)

$$\Pi_{ik}=\rho u_i u_k-\sigma_{ik}=\rho u_i u_k+p\delta_{ik}-\sigma'_{ik}. \tag{2.1.9}$$

The tensor

$$\sigma_{ik}=-p\delta_{ik}+\sigma'_{ik} \tag{2.1.10}$$

is called the (complete) stress tensor, the tensor

$$\sigma'_{ik}=2\nu\rho(e_{ik}-\tfrac{1}{3}\delta_{ik}\Delta),$$

where $\Delta=e_{ii}=\nabla\cdot\boldsymbol{u}$, is called the 'viscous' (or deviatoric) stress tensor, and the tensor

$$e_{ik}=\frac{1}{2}\left(\frac{\partial u_i}{\partial x_k}+\frac{\partial u_k}{\partial x_i}\right) \tag{2.1.11}$$

is called the strain-rate tensor. Here p is the pressure, ν is the kinematic viscosity and the Kronecker delta is defined by

$$\delta_{ik}=\begin{cases}1 & (i=k),\\ 0 & (i\neq k).\end{cases}$$

The flux of momentum through a surface element $\mathrm{d}\boldsymbol{S}=\boldsymbol{n}\,\mathrm{d}S$ fixed in space, where $\boldsymbol{n}$ is a unit vector normal to the surface $\mathrm{d}S$, is given by $n_k\Pi_{ik}\,\mathrm{d}S$. Thus the transport of momentum through the surface fixed in space takes place by means of convective transport of fluid mass ($\rho n_k u_k u_i$), the action of pressure forces ($n_k\delta_{ik}p=n_i p$) and the action of viscous forces ($-n_k\sigma'_{ik}$). The flux of momentum through the surface is, by definition, the force acting on the surface. Hence the ith component of the total force, by which a fluid in the volume V fixed in space acts through its surface S on the surrounding fluid, is

$$\int_S \Pi_{ik}\,\mathrm{d}S_k = \int_S \Pi_{ik} n_k\,\mathrm{d}S$$

(recall that $\boldsymbol{n}$ is directed outward from the surface S).

By definition, the momentum of unit volume of fluid is a vector $\rho\boldsymbol{u} = (\rho u_1, \rho u_2, \rho u_3)$. The conservation of momentum for the volume V fixed in space can be written, similarly to (2.1.3), in the integral form

$$\frac{\mathrm{d}}{\mathrm{d}t}\int_V \rho u_i\,\mathrm{d}V = -\int_S \Pi_{ik}\,\mathrm{d}S_k. \tag{2.1.12}$$

In fact, (2.1.12) represents three independent conditions for the three components of momentum, and this expression is Newton's second law of motion applied to the mass of fluid confined in the volume V. On the left-hand side of (2.1.12) we have the rate of change of momentum of the fluid confined in the volume V, while the right-hand side represents the sum of all external forces acting on this volume through its surface S.

Transforming the surface integral to a volume integral and taking into consideration the fact that (2.1.12) holds for any volume, we finally obtain a differential equation representing the conservation of momentum in tensor notation:

$$\frac{\partial}{\partial t}\rho u_i = -\frac{\partial}{\partial x_k}\Pi_{ik}. \tag{2.1.13}$$

The term on the right-hand side is the divergence, with minus sign, of the tensor Π_{ik}, and a summation over repeated indices is implied, as usual.

Inserting the expression for the tensor Π_{ik} into (2.1.13) and using (2.1.7) and (2.1.8), the equation (2.1.13) for an incompressible fluid can be written in the more common (pseudo)vector notation as

$$\frac{\partial \boldsymbol{u}}{\partial t} + (\boldsymbol{u}\cdot\nabla)\boldsymbol{u} = -\frac{1}{\rho}\nabla p + \nu\nabla^2\boldsymbol{u}, \tag{2.1.14}$$

where

$$\nabla^2 = \nabla\cdot\nabla = \frac{\partial^2}{\partial x_1^2} + \frac{\partial^2}{\partial x_2^2} + \frac{\partial^2}{\partial x_3^2}$$

is a Laplace operator. This relation is the momentum conservation law for a viscous incompressible fluid and is called the Navier–Stokes equation. For a fluid with homogeneous density (2.1.8) and the three components of (2.1.14) include four unknown functions: the three components of velocity u_1, u_2, u_3, and the pressure p (the density $\rho = \text{const}$ is assumed to be known). Thus (2.1.8) and (2.1.14) give a complete set of equations determining the fluid motion. To specify a particular problem, one must add to these equations the appropriate initial and boundary conditions.

To present (2.1.14) in invariant vector notation, suitable for any coordinate system, one must use the identities

$$(\boldsymbol{u}\cdot \nabla)\boldsymbol{u} = \nabla \tfrac{1}{2}\boldsymbol{u}^2 - \boldsymbol{u} \times (\nabla \times \boldsymbol{u}) \equiv \operatorname{grad} \tfrac{1}{2}|\boldsymbol{u}|^2 - \boldsymbol{u} \times \operatorname{curl} \boldsymbol{u},$$

$$\nabla^2 \boldsymbol{u} = \nabla(\nabla\cdot \boldsymbol{u}) - \nabla \times (\nabla \times \boldsymbol{u}) \equiv \operatorname{grad} \operatorname{div} \boldsymbol{u} - \operatorname{curl} \operatorname{curl} \boldsymbol{u}.$$

To calculate the cross-products: $\boldsymbol{u} \times \boldsymbol{v}$, $\nabla \times \boldsymbol{u} \equiv \operatorname{curl} \boldsymbol{u}$ or more complicated terms, it is useful to introduce the isotropic tensor of third order. This is called the alternating tensor, and is defined by

$$\varepsilon_{ijk} = \begin{cases} 1 & \text{if } ijk = 123, 231, 312 \text{ (cyclic order)}, \\ 0 & \text{if any two indices are equal}, \\ -1 & \text{if } ijk = 321, 213, 132 \text{ (anticyclic order)}. \end{cases}$$

Using this tensor, the ith component of the cross-product can be written as

$$(\boldsymbol{u} \times \boldsymbol{v})_i = \varepsilon_{ijk} u_j v_k, \qquad (\nabla \times \boldsymbol{u})_i = \varepsilon_{ijk} \nabla_j u_k.$$

After summation we have $(\boldsymbol{u} \times \boldsymbol{v})_1 = u_2 v_3 - u_3 v_2$, and so on; $(\nabla \times \boldsymbol{u})_1 = \nabla_2 u_3 - \nabla_3 u_2$ and so on.

To calculate the double cross-roduct $\nabla \times (\nabla \times \boldsymbol{u})$, the well-known relation

$$\varepsilon_{ijk}\varepsilon_{klm} = \varepsilon_{kij}\varepsilon_{klm} = \delta_{il}\delta_{jm} - \delta_{im}\delta_{jl}$$

is used:

$$\begin{aligned} [\nabla \times (\nabla \times \boldsymbol{u})]_i &= \varepsilon_{ijk} \nabla_j \varepsilon_{klm} \nabla_l u_m = (\delta_{il}\delta_{jm} - \delta_{im}\delta_{jl}) \nabla_j \nabla_l u_m \\ &= \nabla_i \nabla_j u_j - \nabla_j \nabla_j u_i \equiv [\nabla(\nabla\cdot \boldsymbol{u}) - \nabla^2 \boldsymbol{u}]_i. \end{aligned}$$

With the use of δ_{ik} and ε_{ijk}, the calculations become formal and simple. For more about this and other frequently used mathematical methods of physics the reader should consult the books by Lee (1962) and Arfken (1985).

2.1.4 Sources and sinks of mass; point sources

In deriving the mass conservation law (2.1.3) it was assumed that there are no external sources or sinks of mass inside the fluid. However, in practice one often meets problems in which these external sources are present, and they must be taken into account. The simplest sink of this kind is a thin tube through which the fluid is sucked with a prescribed mass flux. In general, sources can move. Consider for simplicity sources fixed in space.

Let us introduce a distribution of sources and sinks. The distribution is characterized by its density $A(\boldsymbol{x}, t)$, which means that in unit volume of fluid a mass ρA is created (if $A > 0$) or removed (if $A < 0$) in unit time. The mass balance for an arbitrary fixed volume V can be expressed in the form

$$\frac{\mathrm{d}}{\mathrm{d}t}\int_V \rho \, \mathrm{d}V = -\int_S \rho \boldsymbol{u} \cdot \mathrm{d}\boldsymbol{S} + \int_V \rho A \, \mathrm{d}V, \tag{2.1.15}$$

where S is again the surface bounding the volume V. Using a similar procedure as in the derivation of (2.1.6), from (2.1.15) we obtain

$$\frac{1}{\rho}\frac{\mathrm{D}\rho}{\mathrm{D}t} + \nabla \cdot \boldsymbol{u} = A.$$

For an incompressible fluid $\mathrm{D}\rho/\mathrm{D}t = 0$, and the above relation becomes

$$\nabla \cdot \boldsymbol{u} = A.$$

The situation when the mass is created in the whole bulk of the fluid is rather unusual. Usually the density A of a distribution of sources is non-zero only in some localized volume ΔV, while it vanishes in the rest of the fluid. If one has no interest in the details of motion inside the volume ΔV and studies the motion only outside it then the useful mathematical idealization of a point source of mass can be introduced. This can be done by means of a limiting procedure.

Consider the simple case when in the fluid of homogeneous density $A = \text{const} \neq 0$ only in the small volume ΔV, while $A = 0$ outside this region. The total mass ρq created by this source is the integral over the volume ΔV:

$$\rho q = \int_{\Delta V} \rho A \, \mathrm{d}V = \rho A \, \Delta V.$$

Decreasing ΔV and increasing A such that the product $A \, \Delta V$ remains constant,

$$\lim_{\Delta V \to 0, \, A \to \infty} A \Delta V = q,$$

we obtain a 'point' source of mass whose intensity is q. In general, q depends on time. Let this source be at a point $\boldsymbol{x} = \boldsymbol{x}_0$. If we integrate (2.1.15) over a volume that does not include the point $\boldsymbol{x}_0$ then the second integral on the right-hand side of (2.1.15) vanishes and the mass balance is, as usual,

$$\boldsymbol{\nabla} \cdot \boldsymbol{u} = 0 \quad (\boldsymbol{x} \neq \boldsymbol{x}_0).$$

If the point $\boldsymbol{x}_0$ belongs to the volume V, we obtain

$$\int_V \rho A \, \mathrm{d}V = \rho q \lim_{\Delta V \to 0} \frac{1}{\Delta V} \int_{\Delta V} \mathrm{d}V = \rho q.$$

The mass balance (2.1.15) for $\rho = \text{const}$ is expressed as

$$\int_S \boldsymbol{u} \cdot \mathrm{d}\boldsymbol{S} = q,$$

where S is an arbitrary surface around the point $\boldsymbol{x}_0$.

In a similar way, more complex localized configurations of sources and sinks can be considered. Suppose for example that a source and a sink of intensities q and $-q$ respectively are at points $\boldsymbol{x}_0 + \frac{1}{2}\delta \boldsymbol{x}$ and $\boldsymbol{x}_0 - \frac{1}{2}\delta \boldsymbol{x}$ respectively. Decreasing the distance δx and increasing the intensity q such that the product $q \, \delta \boldsymbol{x}$ tends to the finite value

$$\boldsymbol{m} = \lim_{\delta x \to 0, \, q \to \infty} q \delta \boldsymbol{x},$$

we obtain a singularity at the point $\boldsymbol{x}_0$ called a dipole mass source. Note that the intensity $\boldsymbol{m}$ of the dipole is a vector quantity, in contrast to the scalar q.

2.1.5 *External forces; mathematical idealization of a localized force*

The forces acting in fluid media can be divided conveniently into two classes, namely external and internal forces. Internal forces are those produced by the fluid itself. The terms in the expression for the momentum flux tensor Π_{ik} represent the main internal forces. These are included automatically in the momentum conservation law (2.1.12). External forces result from the action of some foreign objects on a fluid. In the case when external forces act on a fluid they should be taken into consideration in the derivation of Newton's second law of motion for the control volume V. External forces like internal forces can be body forces or surface forces. Body forces are distributed throughout the mass of fluid, and result mostly from the action of some external force field on the fluid medium. Body forces are expressed by their density distribution per unit mass ($\boldsymbol{f}$) or per unit volume ($\rho \boldsymbol{f}$) of fluid. The vector $\boldsymbol{f}$ is in general a function of the position vector $\boldsymbol{x}$ and time t, and has the dimensions of acceleration.

Let us add the term $\int_V \rho f_i \, \mathrm{d}V$, expressing the action of external body forces on the fluid in the fixed volume V, to the right-hand side of (2.1.12). We then obtain

$$\frac{\partial \boldsymbol{u}}{\partial t} + (\boldsymbol{u} \cdot \nabla)\boldsymbol{u} = -\frac{1}{\rho} \nabla p + \nu \nabla^2 \boldsymbol{u} + \boldsymbol{f}, \tag{2.1.16}$$

which is called the equation of motion with external body forces. This equation holds for all points of a volume filled with fluid except perhaps some points where there are singularities in the distribution of $\boldsymbol{f}$.

Consider for example the case of a fluid in a uniform gravitational field, with the acceleration due to gravity being a constant vector $\boldsymbol{g} = (0, 0, g)$ in the whole volume of fluid. Then $\boldsymbol{f} = \boldsymbol{g}$, and for a fluid with homogeneous density ($\rho = \text{const}$) the vector $\rho \boldsymbol{g}$ can be represented as the gradient of a scalar function (scalar potential) in the form

$$\rho \boldsymbol{g} = \nabla(\rho g x_3).$$

One can then introduce the so-called reduced pressure P in the form

$$P = p - \rho g x_3,$$

so that the equation of motion (2.1.16), after substitution of P, will not contain the gravitational force explicitly. Thus in a uniform gravitational field the reduced pressure P plays the same role as the pressure p for a weightless fluid. It must be emphasized here that this is the case only when the density is homogeneous. In an inhomogeneous fluid, where density depends on at least the vertical coordinate $\rho = \rho(x_3)$, the term $\rho \boldsymbol{g}$ cannot be expressed as the gradient of a scalar function; it remains in the equation of motion and plays a significant role in the dynamics of fluids with inhomogeneous density (section 2.3).

The second, more complicated, example where a body force appears in the equation of motion relates to the case when the equation of motion is written in a non-inertial frame of reference. Consider a frame rotating at constant angular velocity $\boldsymbol{\Omega} = (0, 0, \Omega)$ with respect to a frame that is fixed in space. In this system there are inertial forces acting on each fluid element. These additional forces can be expressed as the sum of two terms: $\boldsymbol{f} = \boldsymbol{f}_1 + \boldsymbol{f}_2$, where $\boldsymbol{f}_1 = -2\boldsymbol{\Omega} \times \boldsymbol{u}$ is the Coriolis force and $\boldsymbol{f}_2 = -\boldsymbol{\Omega} \times (\boldsymbol{\Omega} \times \boldsymbol{x})$ is the centripetal force.

In the particular case where the fluid has homogeneous density ($\rho = \text{const}$) and the flow is planar $\boldsymbol{u} = (u_1, u_2, 0)$ and independent of the coordinate x_3, the centripetal force $\boldsymbol{f}_2$ can be represented as the gradient of a scalar potential:

$$\boldsymbol{f}_2 = \boldsymbol{x}\Omega^2 = \frac{1}{2\rho} \nabla(\rho |\boldsymbol{x}|^2 \Omega^2).$$

One can then define a reduced pressure as

$$P = p - \tfrac{1}{2}\rho |\boldsymbol{x}|^2 \Omega^2,$$

so that the centripetal force can be excluded from the equation of motion. In contrast, the Coriolis force

$$\boldsymbol{f}_1 = (2\Omega u_2, \ -2\Omega u_1, 0)$$

has no potential and remains in the equation of motion (section 3.3).

In both examples considered here the external body forces act inside the whole bulk of the fluid.

Consider now external forces that are localized in space. A localized external force is one that acts on the fluid only inside a region that is localized and small compared with the whole bulk of the fluid. In general, this small region ΔV can move.

For simplicity, let the volume ΔV be fixed in space. If one is interested only in the motion of fluid outside the volume ΔV, it is convenient to introduce a 'point' source of momentum, similar to the 'point' mass source considered previously. Let $\boldsymbol{f} = \text{const} \neq 0$ inside the volume ΔV and $\boldsymbol{f} = 0$ in the rest of the fluid. The total force $\rho \boldsymbol{F}$ acting on the fluid in the volume ΔV is the volume integral

$$\rho \boldsymbol{F} = \int_{\Delta V} \rho \boldsymbol{f} \, \mathrm{d}V = \rho \boldsymbol{f} \, \Delta V.$$

Decreasing ΔV and increasing the magnitude of $\boldsymbol{f}$ such that $\boldsymbol{F}$ remains constant,

$$\lim_{\Delta V \to 0, \, f \to \infty} \rho \boldsymbol{f} \, \Delta V = \rho \boldsymbol{F},$$

we obtain a finite total force $\rho \boldsymbol{F}$ that acts on the fluid over an infinitesimal volume, i.e. at a point. In general, the magnitude of the force can depend on time. Thus we get a useful mathematical idealization of a localized external force: the 'point' force, which can be considered as a source of momentum. By definition, this source imparts to the fluid an impulse $\rho \boldsymbol{F}$ per unit time in the direction of the vector $\boldsymbol{F}$. In spite of the significant simplifications involved – replacing the finite volume ΔV by a point and replacing the real distribution of forces by only one force – this mathematical idealization turns out to be very useful in the analytical treatment of different problems.

The differential equation of motion for a localized source of momentum can be derived as usual by considering the integral condition of momentum balance in a control volume V:

$$\frac{\mathrm{d}}{\mathrm{d}t} \int_V \rho u_i \, \mathrm{d}V = - \int_S \Pi_{ik} \, \mathrm{d}S_k + \int_V \rho f_i \, \mathrm{d}V. \qquad (2.1.17)$$

If the control volume does not include the point $\boldsymbol{x} = \boldsymbol{x}_0$ where the point force is applied then the last integral in (2.1.17) vanishes and the equation of motion becomes

$$\frac{\partial \boldsymbol{u}}{\partial t} + (\boldsymbol{u} \cdot \nabla) \boldsymbol{u} = -\frac{1}{\rho} \nabla p + \nu \nabla^2 \boldsymbol{u} \quad (\boldsymbol{x} \neq \boldsymbol{x}_0). \qquad (2.1.18)$$

This equation holds at all points of space except $\boldsymbol{x}_0$. If, in contrast, the point $\boldsymbol{x}_0$ is included in the control volume then one can write

$$\int_V \rho \boldsymbol{f}\,\mathrm{d}V = \rho \boldsymbol{F} \lim_{\Delta V \to 0} \frac{1}{\Delta V} \int_{\Delta V} \mathrm{d}V = \rho \boldsymbol{F}$$

and the momentum balance (2.1.17) becomes

$$\frac{\mathrm{d}}{\mathrm{d}t} \int_V \rho u_i \,\mathrm{d}V + \int_S \Pi_{ik}\,\mathrm{d}S_k = \rho F_i, \qquad (2.1.19)$$

where S is the arbitrary surface bounding the volume V and including the point $\boldsymbol{x}_0$. The integral condition (2.1.19) can be considered as the condition of non-triviality of particular solutions of the equation of motion. This condition must be used to connect the characteristics of a particular flow with the intensity of the forcing.

Thus in those cases where the real distribution of external forces can be replaced by an idealized point force the problem is essentially simplified: one has to solve the homogeneous equation (2.1.18) rather than the inhomogeneous equation (2.1.16).

Using the point force idealization it is possible to deal with more complex configurations of localized forcing. Consider for example two point sources of momentum that apply to the fluid equal forces ρF in opposite directions. The sources are at points a distance ε from one another. It is clear that the total force applied to the fluid is zero. Decreasing ε and increasing F such that the product εF remains constant,

$$\lim_{\varepsilon \to 0,\, F \to \infty} \varepsilon F = Q,$$

we obtain a point source of motion of intensity Q, which can be considered as a force dipole. Since we have chosen a simple geometry, the intensity of this source is characterized by the single scalar quantity Q. In general, a force dipole is determined by a second-order tensor Q_{ik}.

2.1.6 Vorticity and the Helmholtz equation·

To characterize a flow field, in addition to the velocity vector $\boldsymbol{u}(\boldsymbol{x}, t)$ another vector quantity $\boldsymbol{\omega}(\boldsymbol{x}, t)$ is frequently used. It is defined by

$$\boldsymbol{\omega} = \nabla \times \boldsymbol{u} \equiv \operatorname{curl} \boldsymbol{u}$$

and is called the vorticity vector. It characterizes the local rotation of a fluid element around its centre of mass, and is equal to the twice the angular velocity of rotation of the fluid element. To see this, consider a simple example. Stokes' theorem states that

$$\int_S (\boldsymbol{\nabla} \times \boldsymbol{u}) \cdot \mathrm{d}\boldsymbol{S} = \oint_C \boldsymbol{u} \cdot \mathrm{d}\boldsymbol{l}, \tag{2.1.20}$$

which says that the line integral of $\boldsymbol{u}$ around a closed curve C is equal to the flux of $\boldsymbol{\nabla} \times \boldsymbol{u}$ through an arbitrary surface S bounded by C. One can apply the above relation to a circle of small radius a around a point $\boldsymbol{x}$. The mean value of the left-hand integral in (2.1.20) is the area of the circle, πa^2, multiplied by $(\boldsymbol{\nabla} \times \boldsymbol{u}) \cdot \boldsymbol{n}$, where $\boldsymbol{n}$ is the unit normal to the surface S. Rewriting (2.1.20) in this way, and using $\boldsymbol{n} \cdot \boldsymbol{n} = 1$, we obtain

$$\tfrac{1}{2}(\boldsymbol{\nabla} \times \boldsymbol{u}) \approx \frac{\boldsymbol{n}}{2\pi a^2} \oint_C \boldsymbol{u} \cdot \mathrm{d}\boldsymbol{l}.$$

The right-hand side of this relation can be considered as a mean value of the tangential velocity component around the circle (whose length is $2\pi a$) divided by the radius a. It is a formal definition of the local angular velocity at a point $\boldsymbol{x}$ directed along the normal $\boldsymbol{n}$. Thus, the vorticity is directly related to the local rotation of fluid elements.

Using the notion of vorticity, irrotational and rotational flows can be defined. Irrotational flows are those fluid motions in which the vorticity vanishes, i.e. $\boldsymbol{\omega}(\boldsymbol{x}, t) = 0$. In contrast, fluid motions in which $\boldsymbol{\omega}(\boldsymbol{x}, t) \neq 0$ are called rotational. Almost all real fluid motions are rotational. However, in many cases it is often possible and very useful to divide the flow field into two parts. The vorticity is concentrated in one of these, and determines the major flow characteristics there. In the other part the vorticity is almost zero, and the flow can be considered as irrotational. There are many reasons why it is convenient to describe rotational flows in terms of the vorticity.

An equation for the rate of change of vorticity is obtained by taking the curl of the equation of motion (2.1.14):

$$\frac{\partial \boldsymbol{\omega}}{\partial t} + \boldsymbol{\nabla} \times (\boldsymbol{u} \cdot \boldsymbol{\nabla})\boldsymbol{u} = \nu \nabla^2 \boldsymbol{\omega}.$$

Note that the pressure is eliminated in this operation because the curl of a gradient vanishes. Since the divergence of a curl also vanishes, the vorticity of any flow must satisfy

$$\nabla \cdot \boldsymbol{\omega} = 0.$$

Using familiar vector identities and writing the term $\nabla \times (\boldsymbol{u} \cdot \nabla)\boldsymbol{u}$ in invariant vector notation, after some algebra we obtain

$$\frac{\partial \boldsymbol{\omega}}{\partial t} + (\boldsymbol{u} \cdot \nabla)\boldsymbol{\omega} = (\boldsymbol{\omega} \cdot \nabla)\boldsymbol{u} + \nu \nabla^2 \boldsymbol{\omega}, \qquad (2.1.21)$$

which is called the Helmholtz equation for the vorticity.

In planar flows when $\boldsymbol{u} = (u_1, u_2, 0)$ and the motion does not depend on the third coordinate x_3 the vorticity vector has only one component $\boldsymbol{\omega} = (0, 0, \omega)$, normal to the plane of motion. In this case the first term on the right-hand side of (2.1.21) vanishes and (2.1.21) becomes

$$\frac{\partial \boldsymbol{\omega}}{\partial t} + (\boldsymbol{u} \cdot \nabla)\boldsymbol{\omega} = \nu \nabla^2 \boldsymbol{\omega}. \qquad (2.1.22)$$

2.1.7 *Stream function and multipole expansion*

The description of flows for which one of the velocity components vanishes and the fluid is incompressible can be considerably simplified by introducing a representation of the velocity such that the mass conservation law (2.1.8) is satisfied identically. For example, for two-dimensional planar flows, when $\boldsymbol{u} = (u_1, u_2, 0)$ and the motion does not depend on the third coordinate x_3, (2.1.8) becomes

$$\frac{\partial u_1}{\partial x_1} + \frac{\partial u_2}{\partial x_2} = 0. \qquad (2.1.23)$$

If a function $\psi(x_1, x_2, t)$ is now defined such that

$$u_1 = \frac{\partial \psi}{\partial x_2}, \qquad u_2 = -\frac{\partial \psi}{\partial x_1} \qquad (2.1.24)$$

then (2.1.23) is automatically satisfied. Thus the number of unknown functions can be decreased from two (u_1, u_2) to one (ψ). The function ψ is called a stream function.

A similar function can be defined for some other cases where the velocity has only two components. Examples include axisymmetric flows and horizontal flows in which $\boldsymbol{u} = (u_1, u_2, 0)$ and u_1, u_2 depend on x_1, x_2, x_3 and t.

For two-dimensional planar flows, when the vorticity has only one component $\boldsymbol{\omega} = (0, 0, \omega)$, substitution of (2.1.24) into the definition of the vorticity gives the kinematic relation between ω and ψ in the form of the Poisson equation

$$\nabla^2 \psi = -\omega. \tag{2.1.25}$$

Assume that a vorticity distribution occupies a finite region S' in a fluid extending to infinity (Figure 2.1) and that the fluid is at rest at infinity. In this case the solution of (2.1.25) can be expressed via the logarithmic potential (e.g. Lee, 1962)

$$\psi(\boldsymbol{x}, t) = -\frac{1}{2\pi} \int_{S'} \omega(\boldsymbol{x}', t) \log |\boldsymbol{x} - \boldsymbol{x}'| \, \mathrm{d}S', \tag{2.1.26}$$

where $\boldsymbol{x} = (x_1, x_2)$. For large distances $|\boldsymbol{x}| \gg |\boldsymbol{x}'|$ the right-hand side of (2.1.26) can be expanded. Expanding $\log |\boldsymbol{x} - \boldsymbol{x}'|$ as a scalar

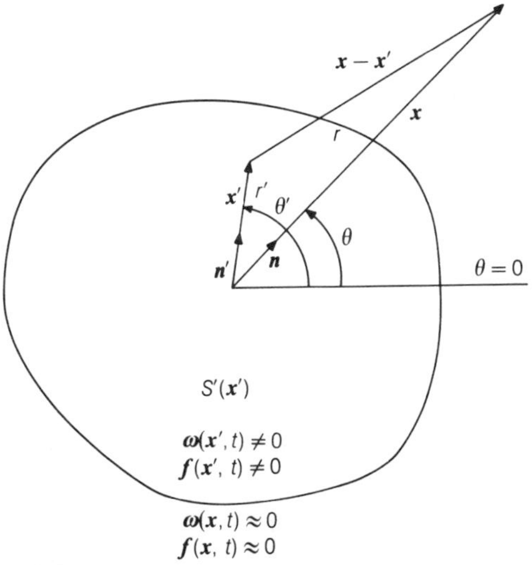

Figure 2.1 *Vorticity distribution and coordinate system.*

function of the coordinate vector $\boldsymbol{x}$ in a Taylor series, we obtain

$$\log|\boldsymbol{x}-\boldsymbol{x}'| = \log|\boldsymbol{x}| - x_i'\frac{\partial \log|\boldsymbol{x}|}{\partial x_i} + \tfrac{1}{2}x_i'x_j'\frac{\partial^2 \log|\boldsymbol{x}|}{\partial x_i \partial x_j} + \cdots$$
$$= \log|\boldsymbol{x}| - \frac{n_i}{|\boldsymbol{x}|}x_i' - \frac{n_i n_j}{2|\boldsymbol{x}|^2}(2x_i'x_j' - |\boldsymbol{x}'|^2\delta_{ij}) + O\left(\frac{1}{|\boldsymbol{x}|^3}\right), \tag{2.1.27}$$

where $n_i = x_i/|\boldsymbol{x}|$, and δ_{ij} is the Kronecker delta. To derive the standard form of the third term in the expansion (2.1.27), we have used the fact that $\log|\boldsymbol{x}|$ satisfies the Laplace equation

$$\nabla^2 \log|\boldsymbol{x}| = \delta_{ij}\frac{\partial^2 \log|\boldsymbol{x}|}{\partial x_i \partial x_j} = 0,$$

so that

$$x_i'x_j'\frac{\partial^2 \log|\boldsymbol{x}|}{\partial x_i \partial x_j} = (x_i'x_j' - \tfrac{1}{2}|\boldsymbol{x}'|^2\delta_{ij})\frac{\partial^2 \log|\boldsymbol{x}|}{\partial x_i \partial x_j}$$
$$= -\frac{n_i n_j}{|\boldsymbol{x}|^2}(2x_i'x_j' - |\boldsymbol{x}'|^2\delta_{ij}).$$

Substitution of (2.1.27) into (2.1.26) gives

$$\psi(\boldsymbol{x}, t) = -\frac{\log|\boldsymbol{x}|}{2\pi}\Gamma + \frac{n_i I_i}{2\pi|\boldsymbol{x}|} + \frac{n_i n_j M_{ij}}{4\pi|\boldsymbol{x}|^2} + O\left(\frac{1}{|\boldsymbol{x}|^3}\right), \tag{2.1.28}$$

where

$$\Gamma = \int_{S'} \omega(\boldsymbol{x}', t)\,\mathrm{d}S', \tag{2.1.29}$$

$$I_i = \int_{S'} x_i'\omega(\boldsymbol{x}', t)\,\mathrm{d}S', \tag{2.1.30}$$

$$M_{ij} = \int_{S'} (2x_i'x_j' - |\boldsymbol{x}'|^2\delta_{ij})\omega(\boldsymbol{x}', t)\,\mathrm{d}S'. \tag{2.1.31}$$

By analogy with the classical multipole expansion from electrodynamics (e.g. Jackson, 1975), the expression (2.1.28) can be called the multipole expansion for the stream function.

To give a geometrical interpretation to each term in (2.1.28), let us rewrite it in a polar coordinate system. Consider for example the third term, denoting it by ψ_3. Referring to Figure 2.1, let θ' and θ be the angles between a fixed axis $\theta = 0$ and the respective coordinate vectors $\boldsymbol{x}'$ and $\boldsymbol{x}$ in a polar coordinate system (r, θ) with origin at $\boldsymbol{x} = 0$. With the help of (2.1.27), the term ψ_3 can then be transformed into

$$\psi_3(\boldsymbol{x}, t) = \frac{1}{4\pi|\boldsymbol{x}|^2} \int_{S'} [2(n_i x_i')(n_j x_j') - |\boldsymbol{x}'|^2 n_i n_j \delta_{ij}] \omega(\boldsymbol{x}', t) \, \mathrm{d}S'$$

$$= \frac{1}{4\pi|\boldsymbol{x}|^2} \int_{S'} [2(\boldsymbol{n}\boldsymbol{n}')^2 |\boldsymbol{x}'|^2 - |\boldsymbol{x}'|^2] \omega(\boldsymbol{x}', t) \, \mathrm{d}S'$$

where $\boldsymbol{n}' = \boldsymbol{x}'/|\boldsymbol{x}'|$ and the identity

$$n_i n_j \delta_{ij} = n_i n_i = |\boldsymbol{n}|^2 = 1$$

has been used. Then, using the relation $(\boldsymbol{n} \cdot \boldsymbol{n}')^2 = \cos^2(\theta - \theta')$, we obtain

$$\psi_3(r, \theta, t) = \frac{1}{4\pi r^2} \cos 2\theta \int_0^{2\pi} \int_0^{\infty} r'^3 \omega(r', \theta', t) \cos 2\theta' \, \mathrm{d}r' \, \mathrm{d}\theta'$$

$$+ \frac{1}{4\pi r^2} \sin 2\theta \int_0^{2\pi} \int_0^{\infty} r'^3 \omega(r', \theta', t) \sin 2\theta' \, \mathrm{d}r' \, \mathrm{d}\theta',$$

where $r = |\boldsymbol{x}|$ and $r' = |\boldsymbol{x}'|$. It is clear that other terms in (2.1.28) can be transformed in a similar way.

Thus for large distances $r \gg r'$ the stream function can be expressed as a series in $\cos n\theta$ and $\sin n\theta$:

$$\psi(r, \theta, t) = -\frac{\log r}{2\pi} \Gamma + \frac{1}{2\pi r}(A_1 \cos\theta + B_1 \sin\theta) + \frac{1}{4\pi r^2}(A_2 \cos 2\theta + B_2 \sin 2\theta) + O\left(\frac{1}{r^3}\right). \tag{2.1.32}$$

where $n = 0, 1, 2, \ldots$. The coefficients Γ, A_n and B_n depend only on time and are equal to the corresponding integral moments of order n of the vorticity distribution.

2.2 Dimensional analysis and self-similarity

In many problems of mathematical physics it is sometimes possible to obtain significant information without solving the particular equations directly, but using only some general principles such as the conservation laws and the dimensional analysis of quantities that are essential for the problem. The most valuable property that can be found in this way is self-similarity of the process under study. This often allows reduction of the number of independent variables and replacement of partial differential equations by ordinary ones, thus essentially simplifying the problem. An important role in elucidating the nature of self-similarity is played by dimensional analysis. To use this one should know some simple basic principles, which are briefly described in this section. For more details the reader should refer to the books by Birkhoff (1960), Sedov (1959) and Barenblatt (1979).

2.2.1 Dimensions of physical quantities

Suppose that some physical quantity a depends on n governing parameters $a_1, \ldots, a_n$:

$$a = f(a_1, \ldots, a_n). \tag{2.2.1}$$

Obviously, although the numerical values of a, $a_1, \ldots, a_n$ depend on a particular system of units, the physical law represented by (2.2.1) must not.

Let us denote by $a_1, \ldots, a_k$ those of the governing parameters whose dimensions are independent. This means that the dimensions of any of these parameters cannot be expressed as a product of powers of dimensions of the other parameters. Then the dimensions of the quantity a and the dimensions of the rest of the arguments $a_{k+1}, \ldots, a_n$ can be expressed in terms of the dimensions of the basic parameters $a_1, \ldots, a_k$:

$$\begin{aligned}
[a] &= [a_1]^p \cdots [a_k]^r, \\
&\;\;\vdots \qquad\quad \vdots \\
[a_{k+1}] &= [a_1]^{p_{k+1}} \cdots [a_k]^{r_{k+1}}, \\
&\;\;\vdots \qquad\quad \vdots \\
[a_n] &= [a_1]^{p_n} \cdots [a_k]^{r_n}.
\end{aligned}$$

Here the square brackets indicate the dimensions of the quantity enclosed. The powers p and r can be obtained by simple counting. Let us define the quantities

$$\Pi = \frac{a}{a_1^p \cdots a_k^r}, \qquad \Pi_1 = \frac{a_{k+1}}{a_1^{p_{k+1}} \cdots a_k^{r_{k+1}}}, \qquad \ldots, \qquad \Pi_{n-k} = \frac{a_n}{a_1^{p_n} \cdots a_k^{r_n}}.$$

These quantities are non-dimensional, so their values are the same for any system of units. Since the relation (2.2.1) corresponds to some physical law that does not depend on the system of units, it can be expressed in the non-dimensional form

$$\Pi = \Phi(\Pi_1, \ldots, \Pi_{n-k}). \tag{2.2.2}$$

Note that this way we have decreased the number of arguments by k.

This result can be summarized by the so-called Π theorem. This states that if some dimensional physical quantity is related to a number of dimensional governing parameters, this relation can be expressed in a non-dimensional form such that the number of non-dimensional arguments is equal to the total number of governing parameters minus the number of parameters whose dimensions are independent.

In some problems the advantage in the number of arguments is crucial. In particular, when $k = n$, (2.2.2) becomes very simple: $\Pi = \text{const}$, and the quantity a is determined to within a constant factor:

$$a = \text{const}\, a_1^p \cdots a_n^r.$$

The unknown constant factor must be determined experimentally or from the full solution of the problem.

2.2.2 *Self-similarity*

If the distributions of quantities characterizing some physical process are similar at different times, such a process is called self-similar. In problems arising in fluid dynamics the quantities characterizing the processes are functions of time t, position vector $\boldsymbol{x} = (x_1, x_2, x_3)$ and also some constant dimensional parameters $a_1, \ldots, a_n$, which determine the particular problem. Thus, in general, we can write

$$a = f(t, x_1, x_2, x_3, a_1, \ldots, a_n). \tag{2.2.3}$$

Assume that, as a result of rewriting (2.2.3) in non-dimensional form, the unknown quantity a turns out to be a function of only the following non-dimensional arguments:

$$\eta_1 = \frac{x_1}{bt^\alpha}, \qquad \eta_2 = \frac{x_2}{bt^\alpha}, \qquad \eta_3 = \frac{x_3}{bt^\alpha}$$

and perhaps of some constant non-dimensional parameters. Here b is a constant, its dimensions are $[b] = LT^{-\alpha}$ and α is a number. This suggests the possible existence of a similarity solution of the problem, and as a result the number of arguments can be reduced from four (x_1, x_2, x_3, t) to three (η_1, η_2, η_3). When the unknown function depends only on one spatial coordinate and time, we obtain an essential simplification of the problem.

The following simple example will help to clarify the meaning of self-similarity. Consider the idealized problem of planar flow of a viscous incompressible fluid induced by a moving plane wall. The wall, which is initially ($t = 0$) at rest, starts to move with a constant speed U_0 along its own plane. Let the wall lie in the (x_1, x_3) plane and move in the positive direction of the x_1 axis. One can expect the flow velocity to have only a single component u_1 directed along the x_1 axis. The unknown distribution of flow velocity depends on four arguments:

$$u_1 = u_1(x_2, t, U_0, \nu). \tag{2.2.4}$$

Two of these four parameters, say U_0 and ν, have independent dimensions

$$[U_0] = LT^{-1}, \qquad [\nu] = L^2T^{-1}.$$

Rewrite the relation (2.2.4) in the non-dimensional form

$$\frac{u_1}{U_0} = f(x_2', t'), \tag{2.2.5}$$

where f is a non-dimensional function, that we are seeking, and $x_2' = x_2 U_0 \nu^{-1}$ and $t' = tU_0^2\nu^{-1}$ are non-dimensional arguments. The Navier–Stokes equation (2.1.14) for this problem becomes

$$\frac{\partial u_1}{\partial t} = \nu \frac{\partial^2 u_1}{\partial x_2^2}. \tag{2.2.6}$$

The mass conservation equation in the form (2.1.8) is satisfied identically. Substituting (2.2.5) into (2.2.6), we obtain the non-dimensional equation for f·

$$\frac{\partial f}{\partial t'} = \frac{\partial^2 f}{\partial x_2'^2}. \tag{2.2.7}$$

Let us now assume that the function f depends on the combination

$$\eta = \frac{x_2'}{2t'^{1/2}} = \frac{x_2}{2(\nu t)^{1/2}}$$

of the arguments x_2' and t', rather than on these arguments separately. Then the partial differential equation (2.2.7) becomes an ordinary differential equation

$$\frac{\mathrm{d}^2 f}{\mathrm{d}\eta^2} + 2\eta \frac{\mathrm{d}f}{\mathrm{d}\eta} = 0. \tag{2.2.8}$$

The boundary conditions can be expressed as

$$f = \begin{cases} 1 & (\eta = 0), \\ 0 & (\eta \to \infty). \end{cases} \tag{2.2.9}$$

The solution of (2.2.8) satisfying (2.2.9) is

$$f = \frac{u_1}{U_0} = \frac{2}{\pi^{1/2}} \int_\eta^\infty \mathrm{e}^{-\eta^2}\,\mathrm{d}\eta, \qquad \eta = \frac{x_2}{2(\nu t)^{1/2}}. \tag{2.2.10}$$

The argument η in (2.2.10) is a so-called similarity variable, and gives the connection between the spatial variable and time. The velocity distributions given by (2.2.10) are affinely related or similar: by changing the length scale of the x_2 axis, we can superpose the velocity distributions for different times. Thus the solution obtained is self-similar. Many other examples of self-similar solutions are considered in the following chapters.

2.3 Density-stratified fluids

A fluid system consisting of horizontal layers of different density is called a density-stratified system, irrespective of whether the density varies in the vertical direction in a step-like manner or continuously. The ocean and atmosphere are the most important examples of fluid systems where the density is not uniform. In the ocean the density of water varies with its salinity and/or temperature. Since in the ocean depths the water is typically cooler than at the surface, the density increases with depth. In the atmosphere the density of air varies with height as a result of temperature and humidity variations.

In contrast to a homogeneous fluid, a density-stratified fluid placed in a uniform gravitational field, characterized by the acceleration vector $\boldsymbol{g} = (0, 0, g)$, exhibits new behaviour. In particular, the gravitational force effectively suppresses motion in a vertical direction (z axis) and makes the flow quasi-two-dimensional, as was demonstrated in section 1.1.

To analyse this important effect quantitatively, the governing equations for a stratified fluid are required. However, before deriving these equations, let us first consider the so-called buoyancy force that appears in a stratified fluid.

2.3.1 *Buoyancy force and buoyancy frequency*

Consider a fluid particle in a density-stratified system. The particle is initially at rest at its equilibrium density level $z = z_0$. Suppose the particle is displaced from its equilibrium level to the level $z_0 + \delta z$, where the density of the surrounding fluid $\rho(z_0 + \delta z)$ differs from the density $\rho(z_0)$ of the particle. At this level the gravitational force on the particle, $g\rho(z_0)$, exceeds the vertical component of the hydrostatic pressure gradient given by

$$\frac{\partial}{\partial z}\int^{z=z_0+\delta z} \rho(z) g \, \mathrm{d}z = \rho(z_0 + \delta z) g.$$

Hence the resulting force on the particle per unit volume is proportional to the density difference $\delta\rho = \rho(z_0) - \rho(z_0 + \delta z)$. This force $g\,\delta\rho$ is called the buoyancy force. It can be directed upwards or downwards, depending on the sign of the displacement δz. The buoyancy force tends to return the particle to its equilibrium level. The falling or rising particle passes its equilibrium level owing to the force of inertia. As a result, the particle oscillates about this level with the so-called buoyancy (or Brunt–Väisälä) frequency, given by

$$\mathscr{N}(z) = \left(\frac{g}{\rho}\frac{\mathrm{d}\rho}{\mathrm{d}z}\right)^{1/2}. \tag{2.3.1}$$

Here z points downwards and $d\rho/dz > 0$. The buoyancy frequency is an important parameter characterizing a stratified fluid. In experiments and in theoretical analysis one often takes the background density profile in the flow under study to have the form

$$\rho(z) = \rho_0 \exp\left(\frac{\mathcal{N}^2 z}{g}\right), \tag{2.3.2}$$

where $\rho_0 = \rho(z = 0)$ is a reference value of the density. For this density profile $\mathcal{N}(z) = \text{const}$.

The depth at which the density increases by a factor e times is $H = g/\mathcal{N}^2$. Taking into consideration that typically $\mathcal{N}$ does not exceed $10^{-2}\,\text{s}^{-1}$ in the ocean and $1\,\text{s}^{-1}$ in laboratory experiments, we obtain the estimates $H = 10^8$ cm (ocean) and 10^4 cm (laboratory). The mean depth of the ocean is about 5×10^5 cm and the depths of fluid layers in laboratory experiments are typically less than 10^2 cm. Hence in both cases the density changes $\delta\rho$ with depth are small ($\delta\rho/\rho \lesssim 3 \times 10^{-2}$), so that the relation (2.3.2) becomes almost linear.

$$\rho(z) \approx \rho_0\left(1 + \frac{\mathcal{N}^2}{g} z\right). \tag{2.3.3}$$

2.3.2 *Equations of motion and the Boussinesq approximation*

The equations governing the dynamics of a fluid with inhomogeneous density in a gravitational field can be derived on the base of the general conservation laws that have been considered in section 2.1. Conservation of mass and momentum are again assumed for inhomogeneous fluid media. These laws can be expressed in integral form for a fixed volume V or for a material, control volume τ. To simplify the resulting differential equations, some additional assumptions, which are realized with very good accuracy in practice, are usually used (e.g. Kamenkovich, 1973).

For example, let the fluid be a solution of salt in water, the temperature being constant. Assume that each component of this mixture can be regarded separately as a continuous medium with its own velocity field. Denote by ρ_w and ρ_s the densities and by $\boldsymbol{u}_w$ and $\boldsymbol{u}_s$ the velocities of the water and salt components respectively. Then we can postulate the conservation of mass of each component for a fixed volume V:

$$\frac{\mathrm{d}}{\mathrm{d}t}\int_V \rho_w \,\mathrm{d}V = -\int_S \rho_w \boldsymbol{u}_w \cdot \mathrm{d}\boldsymbol{S},$$

$$\frac{\mathrm{d}}{\mathrm{d}t}\int_V \rho_s \,\mathrm{d}V = -\int_S \rho_s \boldsymbol{u}_s \cdot \mathrm{d}\boldsymbol{S},$$

where S is the bounding surface of the volume V. Since V is arbitrary, we obtain the conservation equations in differential form:

$$\frac{\partial \rho_w}{\partial t} + \nabla\cdot(\rho_w \boldsymbol{u}_w) = 0, \tag{2.3.4}$$

$$\frac{\partial \rho_s}{\partial t} + \nabla\cdot(\rho_s \boldsymbol{u}_s) = 0. \tag{2.3.5}$$

Let us define the velocity $\boldsymbol{u}$ as the velocity of the centre of mass of particles of salt water

$$\boldsymbol{u} = \frac{\rho_w \boldsymbol{u}_w + \rho_s \boldsymbol{u}_s}{\rho_w + \rho_s}.$$

Taking the sum of (2.3.4) and (2.3.5), we obtain the mass conservation equation for the salt water:

$$\frac{1}{\rho}\frac{\mathrm{D}\rho}{\mathrm{D}t} + \nabla\cdot\boldsymbol{u} = 0, \tag{2.3.6}$$

where $\rho = \rho_w + \rho_s$ is the density of the salt water.

Introducing the vectors

$$\boldsymbol{I}_w = \rho_w(\boldsymbol{u}_w - \boldsymbol{u}),$$
$$\boldsymbol{I}_s = \rho_s(\boldsymbol{u}_s - \boldsymbol{u}),$$

which are the fluxes (per unit area) of water and salt due to diffusion of each component, we can rewrite (2.3.4) and (2.3.5) in the form

$$\frac{\partial \rho_w}{\partial t} + \nabla\cdot(\rho_w \boldsymbol{u} + \boldsymbol{I}_w) = 0, \tag{2.3.7}$$

$$\frac{\partial \rho_s}{\partial t} + \nabla\cdot(\rho_s \boldsymbol{u} + \boldsymbol{I}_s) = 0. \tag{2.3.8}$$

Note that, using the definition of $\boldsymbol{u}$, we immediately obtain

$$\boldsymbol{I}_w + \boldsymbol{I}_s = 0.$$

From the definition (1.1.1) of the salinity s it follows that $\rho_s = \rho s$. Hence, using (2.3.6), we can rewrite (2.3.8) in terms of the salinity as

$$\rho \frac{\mathrm{D}s}{\mathrm{D}t} = -\nabla \cdot \boldsymbol{I}_s. \tag{2.3.9}$$

The diffusive flux of salt $\boldsymbol{I}_s$ is directed against the gradient of salinity ∇s, and thus salt is transported from regions with high concentration to regions with low concentration. This flux is usually expressed in the form (e.g. Batchelor, 1967)

$$\boldsymbol{I}_s = -\rho k \nabla s,$$

where k is the diffusivity of salt in water ($k = 1.1 \times 10^{-5}\,\mathrm{cm}^{-2}\,\mathrm{s}^{-1}$ at $T = 15°\mathrm{C}$). Although k depends on the local characteristics of the medium, the product ρk turns out to be nearly constant, and the condition of conservation of mass of salt (2.3.9) becomes, with good accuracy,

$$\frac{\mathrm{D}s}{\mathrm{D}t} = k \nabla^2 s. \tag{2.3.10}$$

In this form (2.3.10) is usually called the diffusion equation for salt.

Consider now the so-called Boussinesq approximation, which allows one to simplify the equations of motion for a great number of stratified flows. Referring for more detail to the paper by Spiegel and Veronis (1960), we recall here only the basis of this approximation.

Consider first the condition (2.3.6) of conservation of mass for salt water. Let the flow field be characterized by a length scale L, a velocity scale U and a density scale $\delta\rho$, so that the velocity and the density vary by U and $\delta\rho$ respectively over a distance of order L. For an incompressible fluid the change in density at a fixed point, $\partial\rho/\partial t$, is determined only by advection of fluid. Thus for the magnitudes of the terms in (2.3.6) we have the estimates

$$\frac{1}{\rho}\frac{\partial\rho}{\partial t} \approx \frac{U}{L}\frac{\delta\rho}{\rho}, \qquad \frac{1}{\rho}(\boldsymbol{u}\cdot\nabla)\rho \approx \frac{U}{L}\frac{\delta\rho}{\rho}, \qquad \nabla\cdot\boldsymbol{u} \approx \frac{U}{L}.$$

Remembering (section 2.3.1) that for most practical cases $\delta\rho/\rho$ is small ($\delta\rho/\rho \lesssim 3 \times 10^{-2}$), we can replace (2.3.6) by the equation

$$\nabla \cdot \boldsymbol{u} = 0, \tag{2.3.11}$$

which is the same as for an incompressible fluid with homogeneous density.

Consider now how the Boussinesq approximation may be applied to the integral momentum equation (2.1.17) for an inhomogeneous incompressible fluid in the gravitational force field $\boldsymbol{g} = (0, 0, g)$. Taking $\boldsymbol{f} = \boldsymbol{g}$ and using the mass conservation law in the forms (2.3.6) and (2.3.11), we rewrite (2.1.17) in the form of a vector differential equation

$$\frac{\partial \boldsymbol{u}}{\partial t} + (\boldsymbol{u} \cdot \nabla)\boldsymbol{u} = -\frac{1}{\rho}\nabla p + \boldsymbol{g} + \nu \nabla^2 \boldsymbol{u}. \tag{2.3.12}$$

Let $\rho = \rho_0 + \delta\rho$ and $p = p_0 + \delta p$, where $\nabla p_0 = \rho_0 \boldsymbol{g}$ and $\rho_0 = \text{const}$ is some reference density ($\delta\rho \ll \rho_0$). Then, neglecting the second order small terms, we have

$$-\frac{1}{\rho}\nabla p + \boldsymbol{g} = -\frac{1}{\rho_0}(\nabla p_0 + \nabla \delta p)\left(1 - \frac{\delta\rho}{\rho_0} + \cdots\right) + \boldsymbol{g} \approx -\frac{1}{\rho_0}\nabla \delta p + \frac{\delta\rho}{\rho_0}\boldsymbol{g}.$$

Returning to the initial variables p and ρ, we obtain

$$-\frac{1}{\rho}\nabla p + \boldsymbol{g} \approx -\frac{1}{\rho_0}\nabla p + \frac{\rho}{\rho_0}\boldsymbol{g}$$

and (2.3.12) becomes

$$\frac{\partial \boldsymbol{u}}{\partial t} + (\boldsymbol{u} \cdot \nabla)\boldsymbol{u} = -\frac{1}{\rho_0}\nabla p + \frac{\rho}{\rho_0}\boldsymbol{g} + \nu \nabla^2 \boldsymbol{u}, \qquad \rho_0 = \text{const}. \tag{2.3.13}$$

To close the system of governing equations, we must add to (2.3.10), (2.3.11) and (2.3.13) a relation between the density and the salinity, which is called the equation of state. For an incompressible fluid at constant temperature we can use, for example, the linear relation (1.2.2).

Thus (1.2.2), (2.3.10), (2.3.11) and (2.3.13) are the full set of equations governing the motion of an incompressible density-stratified fluid in the Boussinesq approximation.

When the background distribution of density is approximately linear and is given, for example, by (2.3.3), where $\mathcal{N} \approx \text{const}$, the governing equations can be presented in a more convenient form for analytical study. Let

$$\rho(\boldsymbol{x}, t) = \rho_b(z) + \tilde{\rho}(\boldsymbol{x}, t),$$
$$p(\boldsymbol{x}, t) = p_b(z) + \tilde{p}(\boldsymbol{x}, t),$$

where $\nabla p_b = \rho_b \boldsymbol{g}$, with $\rho_b(z)$ the background vertical density distribution when the fluid is at rest, and $\tilde{\rho}(\boldsymbol{x}, t)$ and $\tilde{p}(\boldsymbol{x}, t)$ are the perturbations caused by motion of the fluid ($\tilde{\rho} \ll \rho_b$).

Inserting (1.2.2) into (2.3.10) and excluding the salinity from consideration, we can present the condition of conservation of mass of salt (2.3.10) in the form

$$\frac{\partial \rho}{\partial t} + (\boldsymbol{u} \cdot \nabla)\rho = k \nabla^2 \rho. \tag{2.3.14}$$

In terms of the perturbations this equation becomes

$$\frac{\partial \tilde{\rho}}{\partial t} + (\boldsymbol{u} \cdot \nabla)\tilde{\rho} + w\frac{\rho_0 N^2}{g} = k \nabla^2 \tilde{\rho}, \tag{2.3.15}$$

where w is the vertical component of the velocity. For the momentum equation (2.3.13) we have

$$\frac{\partial \boldsymbol{u}}{\partial t} + (\boldsymbol{u} \cdot \nabla)\boldsymbol{u} = -\frac{1}{\rho_0} \nabla \tilde{p} + \frac{\tilde{\rho}}{\rho_0} \boldsymbol{g} + \nu \nabla^2 \boldsymbol{u}. \tag{2.3.16}$$

Adding to (2.3.15) and (2.3.16) the mass conservation equation

$$\nabla \cdot \boldsymbol{u} = 0, \tag{2.3.17}$$

we obtain the set of governing equations (2.3.15)–(2.3.17) for a linearly stratified fluid, written in terms of the perturbations $\tilde{\rho}$ and $\tilde{p}$.

2.3.3 Quasi-two-dimensional flows: scaling analysis

The results of experiments (discussed in section 2.1) demonstrate that the buoyancy force rapidly suppresses motion in the vertical direction and makes the flow horizontal. Now that we have a full set of governing equations, let us consider analytically this specific property of the flows induced in a stratified fluid under the horizontal action of some localized momentum source (Voropayev and Afanasyev, 1991). The full nonlinear unsteady problem is too complicated for analysis. For simplicity consider first the steady flow of a viscous incompressible linearly stratified fluid extending to infinity. The localized source is at the point $(x, y, z) = (0, 0, 0)$ and applies a force $\rho_0 J = \text{const}$ in the direction along the x axis, with the dimensions of J being $L^4 T^{-2}$. This flow can be considered as a steady submerged horizontal jet.

The main purpose of the following analysis is to understand how the flow becomes quasi-two-dimensional and at what values of the external governing parameters this happens.

In spite of the fact that the flow has no length scale or velocity scale of its own, we can introduce such scales artificially, assuming that they should not enter into the solution. This is a standard procedure widely used for this kind of flow (e.g. Schlichting, 1979). Take the longitudinal velocity U_0 on the axis of the jet as the velocity scale and the vertical scale H of the jet as the length scale. In fact, the intensity of forcing J, the kinematic viscosity ν and the buoyancy frequency $\mathcal{N}$ are the only governing parameters appearing in the problem, so these parameters must determine finally all the major flow characteristics, including U_0 and H.

Using the introduced scales, we can rewrite the governing equations (2.3.15)–(2.3.17) in the non-dimensional form

$$u'\frac{\partial u'}{\partial x'} + v'\frac{\partial u'}{\partial y'} + w'\frac{\partial u'}{\partial z'} = -\frac{\partial p'}{\partial x'} + Re^{-1}\left(\frac{\partial^2 u'}{\partial x'^2} + \frac{\partial^2 u'}{\partial y'^2} + \frac{\partial^2 u'}{\partial z'^2}\right), \tag{2.3.18}$$

$$u'\frac{\partial v'}{\partial x'} + v'\frac{\partial v'}{\partial y'} + w'\frac{\partial v'}{\partial z'} = -\frac{\partial p'}{\partial y'} + Re^{-1}\left(\frac{\partial^2 v'}{\partial x'^2} + \frac{\partial^2 v'}{\partial y'^2} + \frac{\partial^2 v'}{\partial z'^2}\right), \tag{2.3.19}$$

$$u'\frac{\partial w'}{\partial x'} + v'\frac{\partial w'}{\partial y'} + w'\frac{\partial w'}{\partial z'} = -\frac{\partial p'}{\partial z'} + Ri\,\rho' + Re^{-1}\left(\frac{\partial^2 w'}{\partial x'^2} + \frac{\partial^2 w'}{\partial y'^2} + \frac{\partial^2 w'}{\partial z'^2}\right), \tag{2.3.20}$$

$$u'\frac{\partial \rho'}{\partial x'} + v'\frac{\partial \rho'}{\partial y'} + w'\frac{\partial \rho'}{\partial z'} + w' = Re^{-1}\,Sc^{-1}\left(\frac{\partial^2 \rho'}{\partial x'^2} + \frac{\partial^2 \rho'}{\partial y'^2} + \frac{\partial^2 \rho'}{\partial z'^2}\right), \tag{2.3.21}$$

$$\frac{\partial u'}{\partial x'} + \frac{\partial v'}{\partial y'} + \frac{\partial w'}{\partial z'} = 0. \tag{2.3.22}$$

Here $(u', v', w') = (uU_0^{-1}, vU_0^{-1}, wU_0^{-1})$ is the non-dimensional velocity vector in non-dimensional Cartesian coordinates

$$(x', y', z') = (xH^{-1}, yH^{-1}, zH^{-1}), \text{ and}$$

$$p' = \tilde{p}(\rho_0 U_0^2)^{-1}$$

$$\rho' = \tilde{\rho}\left(H\frac{\mathrm{d}\rho}{\mathrm{d}z}\right)^{-1} = \tilde{\rho}\left(H\rho_0\frac{\mathcal{N}^2}{g}\right)^{-1}$$

are the non-dimensional pressure and density perturbations. The Reynolds number

$$Re = HU_0/\nu$$

represents the ratio of convective to viscous terms; the Richardson number

$$Ri = H^2\mathcal{N}^2/U_0^2$$

represents the ratio of buoyancy to the inertial forces; and, the Schmidt number

$$Sc = \nu/k$$

represents the ratio of the diffusivity of momentum to the diffusivity of salt.

To simplify the set of equations (2.3.18)–(2.3.22) and obtain some estimates, we recall some well-known ideas from Prandtl's boundary layer theory (e.g. Schlichting, 1979). According to this theory, the boundary layer is a thin layer near the surface of a solid body where the viscosity is significant in flow dynamics at high Reynolds number. A characteristic feature of the boundary layer is that the parameters of the flow vary significantly across the layer while the variations along the layer are relatively small. Experiments demonstrate (Figure 1.8) that the jet flow is localized in a thin cone along the x axis; thus we can consider a steady jet as a free (i.e. without solid boundaries) three-dimensional boundary layer. According to the standard procedure used in the boundary layer theory, we assume that for $Re \gg 1$ the convective terms in (2.3.18)–(2.3.20) are balanced by the terms expressing the viscous diffusion of momentum across the boundary layer. From (2.3.18) we obtain the asymptotic estimates

$$x' = Re\,x^*, \qquad y' = y^*, \qquad z' = z^*,$$

$$u' = u^*, \qquad v' = Re^{-1}v^*, \qquad w' = \alpha\,Re^{-1}w^*,$$

where the asterisks indicate asymptotic variables of order unity. Obviously, w' does not exceed v'; hence $\alpha < 1$ (a more precise estimate of α is given below). From (2.3.19) we obtain the estimate

$$\frac{\partial p'}{\partial y'} = Re^{-2}\frac{\partial p^*}{\partial y^*}.$$

Thus the pressure difference $\delta p'$ across the jet is small, $\delta p' \approx Re^{-2}$, and hence

$$\frac{\partial p'}{\partial z'} = Re^{-2}\frac{\partial p^*}{\partial z^*}.$$

In (2.3.20) the term $Ri\,\rho'$ must not exceed the largest term $\partial p'/\partial z'$; hence we obtain

$$\rho' = Re^{-2}\,Ri^{-1}\rho^*.$$

The Schmidt number Sc for an aqueous solution of salt is large ($Sc \approx 10^3$), and from (2.3.21) we have $\alpha = Re^{-2}\,Ri^{-1}$, so that

$$w' = Re^{-3}\,Ri^{-1}w^*.$$

In boundary layer theory the longitudinal pressure gradient $\partial p'/\partial x'$ is determined by the known gradient in the irrotational flow outside the boundary layer. In our problem the ambient fluid is at rest. Thus we have

$$\frac{\partial p'}{\partial x'} = 0.$$

Substituting the estimates obtained into (2.3.18)–(2.3.22) and neglecting terms that are small when $Re \gg 1$ and $Re\,Ri > 1$, we obtain the reduced asymptotic equations

$$u^*\frac{\partial u^*}{\partial x^*} + v^*\frac{\partial u^*}{\partial y^*} = \frac{\partial^2 u^*}{\partial y^{*2}} + \frac{\partial^2 u^*}{\partial z^{*2}}, \tag{2.3.23}$$

$$\frac{\partial u^*}{\partial x^*} + \frac{\partial v^*}{\partial y^*} = 0, \qquad p' = \rho' = w' = 0. \tag{2.3.24}$$

The boundary conditions are

$$\left.\begin{aligned} &\frac{\partial u^*}{\partial y^*}=0, \quad v^*=0 \quad (y^*=0), \\ &\frac{\partial u^*}{\partial z^*}=0 \qquad\qquad (z^*=0), \\ &u^*=0 \qquad\qquad (y^{*2}+z^{*2}\to\infty). \end{aligned}\right\} \tag{2.3.25}$$

The relation between the momentum imparted by the source per unit time and the total momentum flux in the direction of the x axis serves as the condition of non-triviality of the solution:

$$\frac{J}{v^2\,Re^2}=\int_{-\infty}^{\infty}\int_{-\infty}^{\infty} u^{*2}\,\mathrm{d}y^*\,\mathrm{d}z^*=1, \tag{2.3.26}$$

where

$$J/v^2=Re^2$$

is the non-dimensional intensity of the source. Note that the above relation gives the connection between the governing parameters J, v and the artificial scales U_0 and H, which are included in Re.

The Richardson number Ri, which expresses the influence of the buoyancy force on the flow, enters neither the resulting governing equations (2.3.23) and (2.3.24) nor the boundary conditions (2.3.25) and (2.3.26). However, it is the buoyancy force that makes the flow quasi-planar. The fluid particles move in horizontal planes, and the transport of momentum between ambient planes takes place by means of viscous diffusion.

The equations (2.3.23) and (2.3.24) and the boundary conditions (2.3.25) and (2.3.26) permit transformation of the independent variables:

$$(x^*, y^*, z^*)\to(X, Y, Z)=\left(x^*, \frac{y^*}{x^*}, \frac{z^*}{x^*}\right).$$

Substitution of these transformed variables into the equations allows one to obtain an estimate of the longitudinal component of velocity in the flow:

$$u(x, y, z)=U(Y, Z)\,Re^2\frac{v}{x}, \tag{2.3.27}$$

where

$$Y = Re\frac{y}{x}, \qquad Z = Re\frac{z}{x}.$$

Thus the velocity distributions at different distances are affinely related. From (2.3.27) we immediately obtain an estimate of the velocity at the jet axis:

$$u(x, 0, 0) \equiv U_0 = A\,Re^2\frac{\nu}{x}, \tag{2.3.28}$$

where

$$A = U(0, 0) = \text{const.}$$

In fact, the results of laboratory measurements of U_0 are in good agreement with the estimate (2.3.28), and give a value of $A = 0.12 \pm 0.02$ (Figure 2.2).

Longitudinal velocity distributions across the jet in the vertical plane ($y = 0$) and the horizontal plane ($z = 0$) at different distances x from the origin are shown in Figure 2.3 (symbols represent measured values). Similar coordinates Y and Z are used to compare the

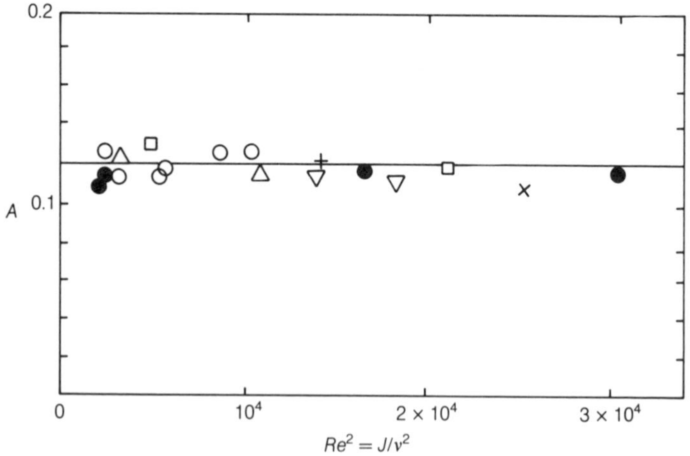

Figure 2.2 *Mean values of A versus non-dimensional intensity of the source* $J/\nu^2 = Re^2$ *for experiments with various* $\mathcal{N}$: +, $0.4\,s^{-1}$; ×, $0.6\,s^{-1}$; ○, $0.7\,s^{-1}$; ∇, $1.0\,s^{-1}$; Δ, $1.1\,s^{-1}$; ●, $1.5\,s^{-1}$; □, $1.6\,s^{-1}$.

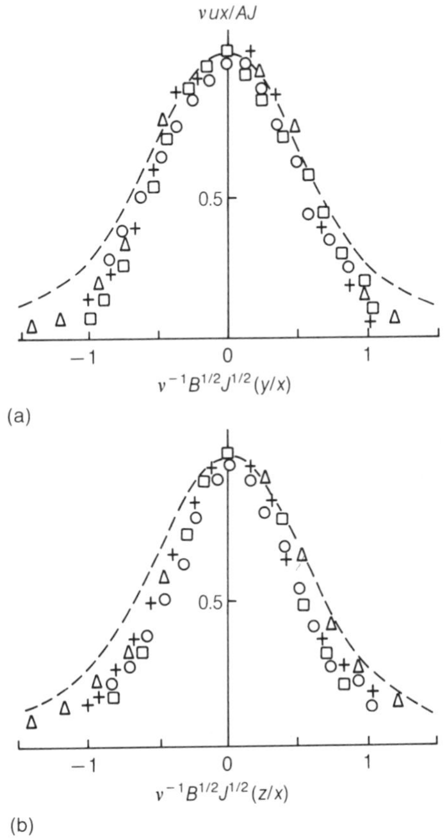

Figure 2.3 *Distributions of the longitudinal velocity in a horizontal jet in a stratified fluid: (a) in the horizontal plane ($z = 0$); (b) in the vertical plane ($y = 0$). The symbols represent data from experiments with $\mathcal{N} = 0.7\,s^{-1}$: Δ, Re = 99, Ri = 0.24, x = 5 cm; □, 69, 0.25, 10 cm; +, 96, 2.27, 15 cm; ○, 50, 14.3, 20 cm. The dashed lines represent (a) Schlichting's distribution (2.3.29) in a homogeneous fluid and (b) the results of three-dimensional numerical simulations at Re = 60, Ri = 9.47, $\mathcal{N} = 1\,s^{-1}$ and x = 5.5 cm.*

measured velocity distributions with the relation (2.3.27). One can see that the measured profiles are indeed affinely related, and do not depend on the stratification parameter *Ri*. The dashed line in Figure 2.3(a) shows the theoretical distribution obtained by Schlichting.

This distribution represents the solution for an axisymmetric jet in a homogeneous fluid (Shlichting, 1933, 1979). The solution was obtained within the boundary layer approximation and has the form

$$u = \frac{AJ\nu^{-1}}{x(1 + BJr^2/\nu^2x^2)^2}, \tag{2.3.29}$$

where

$$r^2 = y^2 + z^2, \qquad A = \frac{3}{8\pi}, \qquad B = \frac{3}{64\pi}.$$

The results of numerical simulations of three-dimensional unsteady flow (Voropayev and Neelov, 1991) are shown by the dashed line in Figure 2.3(b). This distribution was taken just behind the front of the jet. One can see that Schlichting's distribution is a good approximation for both vertical and horizontal distributions of the velocity in a jet in a stratified fluid. It should be noted that this result is not obvious. On streak pictures (Figure 1.8) one can see that the flow is different in the horizontal and vertical planes. The vertical velocity is practically absent, except in the immediate vicinity of the orifice, but the horizontal (lateral) velocity is present. From these pictures alone one might expect that the velocity distributions in the horizontal and vertical planes will be different, but measurements and numerical simulations demonstrate that this is not so.

Substituting (2.3.28) into the definition of the Reynolds number $Re = U_0H/\nu$, we immediately obtain the estimate

$$H = \frac{x}{A\,Re} = \frac{x\nu}{AJ^{1/2}}.$$

Thus the width of the jet decreases as the intensity of the source, which is proportional to the Reynolds number, increases. For the Richardson number, we find

$$Ri = \frac{H^2\mathcal{N}^2}{U_0^2} = \frac{\mathcal{N}^2x^4}{A^4\,Re^6\nu^2} \propto x^4.$$

Thus for the flows considered here the Richardson number increases rapidly with distance x from the origin, in contrast to intrusions of gravity current type, which travel at constant Richardson number (Simpson, 1987). The Reynolds number of the flow considered here remains constant along the flow as a consequence of momentum conservation (2.3.26). The vertical component of velocity decays with distance x as $w \propto Ri^{-1} \propto x^{-4}$, so that the flow rapidly becomes horizontal. Other flows of this type also become quasi-two-dimensional at the considered asymptotic stage – only the transitional stages may be different. These transitional stages can be characterized by the power n in the relation $Ri \propto x^n$. For example, $n = 3$ for an impulsive vortex dipole (section 5.2), while $n = 4$ for a starting vortex dipole at the front of a developing jet (section 5.1).

3

Vortex multipoles

In this chapter the motion of an incompressible fluid of homogeneous density induced by a localized distribution of vorticity is considered. In an ideal fluid the vorticity distribution is set up artificially. In a viscous fluid the vorticity distribution is created by the action of external forcing. The fluid extends to infinity and is at rest there.

In most of the following analysis a standard stream function ψ is used. However, in some cases it is useful to introduce a vector potential $\boldsymbol{B}$. If there are no volume sources in the fluid, the velocity can be expressed purely as the curl of this vector potential:

$$\boldsymbol{u} = \nabla \times \boldsymbol{B}. \tag{3.0.1}$$

The mass conservation equation is satisfied identically, since

$$\nabla \cdot \boldsymbol{u} = \nabla \cdot (\nabla \times \boldsymbol{B}) = 0.$$

The divergence of $\boldsymbol{B}$ is arbitrary. The convenient choice of the Coulomb gauge $\nabla \cdot \boldsymbol{B} = 0$ leads, with the aid of a vector identity

$$\nabla \times (\nabla \times \boldsymbol{B}) = \nabla(\nabla \cdot \boldsymbol{B}) - (\nabla \cdot \nabla)\boldsymbol{B},$$

to the vector Poisson equation

$$\nabla^2 \boldsymbol{B} = -\boldsymbol{\omega}, \tag{3.0.2}$$

where $\boldsymbol{\omega} = \nabla \times \boldsymbol{u}$ is, as usual, the vorticity vector.

For two-dimensional planar motions when $\boldsymbol{u} = (u_1, u_2, 0)$ and the flow characteristics do not depend on the coordinate x_3, the vector potential can be expressed in the simple form

$$\boldsymbol{B} = \boldsymbol{k}\psi = (0, 0, \psi),$$

where ψ is a stream function and $\boldsymbol{k}$ is a unit vector along the x_3 axis, which is perpendicular to the plane of motion. In this case the vorticity has only one component, $\boldsymbol{\omega} = (0, 0, \omega)$, and (3.0.2) becomes

$$\boldsymbol{k}\,\nabla^2\psi = -\,\omega\boldsymbol{k}. \tag{3.0.3}$$

If the flow is irrotational and there are no internal boundaries in the fluid, the stream function can be found from the Laplace equation

$$\nabla^2\psi = 0. \tag{3.0.4}$$

When the fluid is at rest at infinity, the only solution of (3.0.4) is the trivial state of rest. Many interesting problems arise when the fluid contains a vortical region. In this case, to obtain the stream function, one has to solve the inhomogeneous equation (3.0.3).

The asymptotic relation (2.1.28) shows that the far-field stream function for the flow induced by a localized vorticity distribution (Figure 2.1) is represented by an expansion in inverse powers of the distance, the coefficients being proportional to the integrals of vorticity (2.1.29), dipole moment of vorticity (2.1.30), quadrupole moment of vorticity (2.1.31) etc. Vortex flows whose dynamics are governed by some of these parameters are considered below. By analogy with classical electrodynamics, these flows are called vortex multipoles. These multipoles are considered first in an ideal fluid, and then viscous effects are taken into consideration.

When an inviscid fluid contains a localized region with known vorticity distribution, one may try to solve the problem analytically if the form of the vortical region is simple enough, e.g. if the region is a point or it has some symmetry. When the vortical region is rather complex, direct numerical calculations (or the so-called contour dynamics method) are required to obtain the induced flow field (e.g. McWilliams and Zabusky, 1982; Dritschel, 1986; Stern, 1987). In an ideal fluid the vorticity is discontinuous at the boundaries of the vortical regions (or vortical patches), although the velocity is continuous. In a real fluid the viscosity smooths the discontinuities, thus significantly affecting the dynamics of the flow. In addition, the regions of vorticity are usually created during the previous evolution of the flow under study, and the viscosity plays an essential role in this process.

3.1 Multipoles in an ideal fluid

In this section steady vortex multipoles induced by point singularities of the vorticity distribution in an ideal fluid are considered. Although these steady solutions are not very useful in practice, because in a viscous fluid diffusion makes a flow of this type essentially unsteady, they still satisfy the full steady Navier–Stokes equation and can serve as initial states in problems connected with unsteady vortex multipoles in a viscous fluid. These steady solutions also help in understanding the flow kinematics.

3.1.1 *The Bernoulli equation*

For steady motion of an incompressible fluid, with the aid of the vector identities

$$\boldsymbol{u} \times (\boldsymbol{\nabla} \times \boldsymbol{u}) = \tfrac{1}{2}\boldsymbol{\nabla}|\boldsymbol{u}|^2 - (\boldsymbol{u} \cdot \boldsymbol{\nabla})\boldsymbol{u},$$
$$\boldsymbol{\nabla} \times (\boldsymbol{\nabla} \times \boldsymbol{u}) = \boldsymbol{\nabla}(\boldsymbol{\nabla} \cdot \boldsymbol{u}) - \nabla^2 \boldsymbol{u} = -\nabla^2 \boldsymbol{u},$$

we can rewrite the equation of motion (2.1.14) as

$$\boldsymbol{\nabla}\left(\tfrac{1}{2}|\boldsymbol{u}|^2 + \frac{p}{\rho}\right) = -\nu(\boldsymbol{\nabla} \times \boldsymbol{\omega}) + \boldsymbol{u} \times \boldsymbol{\omega}. \tag{3.1.1}$$

In a region where the flow is irrotational the right-hand side vanishes and (3.1.1) becomes

$$\boldsymbol{\nabla} H = 0, \tag{3.1.2}$$

where $H = \tfrac{1}{2}|\boldsymbol{u}|^2 + p/\rho$. Integrating (3.1.2), we obtain the so-called Bernoulli equation

$$H = \tfrac{1}{2}|\boldsymbol{u}|^2 + \frac{p}{\rho} = \text{const.}$$

Here the constant is the same for the whole region of irrotational flow.

Thus, to determine the properties of irrotational flow, one has to solve the linear Laplace equation (3.0.4) with appropriate conditions at the boundary of the region of irrotational flow, while the equation of motion in the form of the Bernoulli equation is used for determining the pressure. Note that this is the case for a viscous fluid as well as for an ideal fluid. The resulting viscous force acting on a fluid element, $-\nu(\boldsymbol{\nabla} \times \boldsymbol{\omega})$, is zero when $\omega = 0$, while the viscous forces themselves may be present.

Consider now a region of ideal fluid ($\nu = 0$) where the flow is vortical. The equation of motion (3.1.1) becomes

$$\nabla H = \boldsymbol{u} \times \boldsymbol{\omega}. \tag{3.1.3}$$

The gradient ∇H is a vector normal to the surface $H = \text{const}$, whereas the right-hand side of (3.1.3) is a vector perpendicular to both $\boldsymbol{u}$ and $\boldsymbol{\omega}$. This gives

$$\boldsymbol{u} \cdot \nabla H = 0, \qquad \boldsymbol{\omega} \cdot \nabla H = 0.$$

It follows that a surface of constant H must contain the streamlines as well as the vortex lines. Thus in an inviscid steady vortical flow H is constant along a streamline; hence H depends only on ψ. Now we can rewrite (3.1.3) in the form

$$\boldsymbol{u} \times \boldsymbol{\omega} = \frac{\mathrm{d}H}{\mathrm{d}\psi} \nabla \psi. \tag{3.1.4}$$

Consider the vorticity balance (2.1.22). For a steady planar flow of inviscid fluid (2.1.22) becomes

$$(\boldsymbol{u} \cdot \nabla)\boldsymbol{\omega} = 0. \tag{3.1.5}$$

This means that the vorticity of a fluid element is conserved along its path. Taking into account that the particle trajectories or path lines and the streamlines are identical in a steady flow, we immediately obtain that the vorticity is constant along streamlines, so that it can be expressed as a function of ψ alone:

$$\omega = f(\psi).$$

Thus for the region of vortical flow (3.0.3) can be written in the form

$$\nabla^2 \psi = -f(\psi), \tag{3.1.6}$$

which determines the distribution of velocity in the steady flow if the function $f(\psi)$ is known.

Comparing the components on each side of the vector equation (3.1.4) and using the fact that $u_1 = \partial\psi/\partial x_2$ and $u_2 = -\partial\psi/\partial x_1$, we can rewrite (3.1.4) as a scalar relation

$$\frac{\mathrm{d}H}{\mathrm{d}\psi} = -\omega = -f(\psi). \tag{3.1.7}$$

Integrating this for the known function $f(\psi)$, one can try to determine the quantity H and hence the pressure. Considering an inviscid fluid, one may take, in principle, an arbitrary distribution of vorticity and then, by solving (3.1.6) and (3.1.7), try to obtain all the flow properties. However, although in principle the mathematical aspects of the problem are clear, the question still remains as to whether the solution obtained is related to a real flow. In practice, the distribution of vorticity in a steady flow is formed during the preceding evolution of the flow, and the viscosity usually plays an essential role during this process. In general, it is not possible to analyse the establishment of a steady flow, and only in some simple cases can the function f be determined. An example is given in section 3.1.3.

3.1.2 Point multipoles

Fundamental solutions ψ_n of the Laplace equation (3.0.4), which for two dimensions can be written in polar coordinates (r, θ) as

$$\nabla^2\psi = \frac{1}{r}\frac{\partial}{\partial r}\left(r\frac{\partial\psi}{\partial r}\right) + \frac{1}{r^2}\frac{\partial^2\psi}{\partial\theta^2} = 0, \tag{3.1.8}$$

are given by

$$\psi_n = \frac{\partial^n(\log r)}{\partial x_i\,\partial x_j\cdots} \quad (n = 0, 1, 2, \ldots)$$

Performing the differentiations, we obtain

$$\psi_0 = \log r, \qquad \psi_n \propto r^{-n}\begin{bmatrix}\cos n\theta\\ \sin n\theta\end{bmatrix} \quad (n = 1, 2, \ldots).$$

Note that functions $r^{2n}\psi_n$ also satisfy (3.1.8). Thus, if we consider an irrotational flow outside some circle of arbitrary finite radius around the origin, we can represent the stream function of the flow as a superposition of independent solutions ψ_n. In describing the flow inside the circle we should use the functions $r^{2n}\psi_n$ with positive powers of n.

Consider for simplicity a small localized vortical region, which can be idealized by a concentrated vortex. In this case the stream function can be represented by (2.1.28) or (2.1.32). Considering different types of singularities at the point where the vortex is concentrated, it is possible to obtain solutions for multipoles of different order.

The simplest multipole is a vortex monopole. A real vortex, with a core within which vorticity is distributed, can be idealized by a concentrated line with a strength equal to the average vorticity in the core multiplied by the core area. Thus a vortex tube is transformed to a vortex line, and the only vortex line in the flow field will be the axis of the vortex. Recall that a vortex line is a curve in the fluid whose tangent at any point gives the direction of the local vorticity, and vortex lines passing through any closed curve form a tubular surface known as a vortex tube. A line vortex is characterized by its strength and by the form of the line.

Consider a single rectilinear line vortex of intensity Γ. In two dimensions this will correspond to a vortex at a point $\boldsymbol{x}'$. A steady irrotational flow field throughout the plane except at the point $\boldsymbol{x}'$ can easily be found from the expansion (2.1.28), where only the first term remains, so that

$$\psi = -\frac{\Gamma}{2\pi}\log|\boldsymbol{x}-\boldsymbol{x}'| = -\frac{\Gamma}{2\pi}\log r, \tag{3.1.9}$$

where $r = |\boldsymbol{x}-\boldsymbol{x}'|$ is the distance from the point vortex. This relation corresponds to the zeroth-order solution ψ_0 of the Laplace equation. It follows from (3.1.9) that

$$u = \frac{1}{r}\frac{\partial\psi}{\partial\theta} = 0, \qquad v = -\frac{\partial\psi}{\partial r} = \frac{\Gamma}{2\pi r}. \tag{3.1.10}$$

Thus fluid particles move in circular paths. The circulation around an arbitrary closed curve L within which lies the point $\boldsymbol{x}'$ is the contour integral

$$\oint_L \boldsymbol{u}\cdot \mathrm{d}\boldsymbol{L} = \int_0^{2\pi} vr\,\mathrm{d}\theta = \Gamma.$$

The vorticity in this flow vanishes everywhere except at the point $\boldsymbol{x}'$, where there is a singularity characterized by the single integral intensity Γ.

Using the same procedure, it is possible to construct more complex singularities. Suppose we have two point vortices of intensities Γ

and $-\Gamma$ at the points $\boldsymbol{x}'+\frac{1}{2}\delta\boldsymbol{x}'$ and $\boldsymbol{x}'-\frac{1}{2}\delta\boldsymbol{x}'$ respectively. These vortices have circulations of the same magnitude Γ but opposite senses of rotation. Now let us decrease the distance $|\delta\boldsymbol{x}'|$ between the vortices and increase their intensity Γ such that the product $\Gamma\delta\boldsymbol{x}'$ remains constant:

$$\boldsymbol{I}=\lim_{\delta|\boldsymbol{x}'|\to 0,\ \Gamma\to\infty}\Gamma\delta\boldsymbol{x}'.$$

As a result, at the point $\boldsymbol{x}'$ we obtain a singularity known as an impulsive vortex dipole of intensity $\boldsymbol{I}$. The irrotational flow field induced by the dipole can be represented as the superposition of the flows induced by each point vortex separately. (Recall that irrotational motion is governed by the linear Laplace equation.) Using (3.1.9), we obtain

$$\begin{aligned}\psi(\boldsymbol{x})&=-\lim_{|\delta\boldsymbol{x}'|\to 0}\frac{\Gamma}{2\pi}(\log|\boldsymbol{x}-\boldsymbol{x}'+\tfrac{1}{2}\delta\boldsymbol{x}'|-\log|\boldsymbol{x}-\boldsymbol{x}'-\tfrac{1}{2}\delta\boldsymbol{x}'|)\\&=-\frac{\boldsymbol{I}}{2\pi}\nabla'\log|\boldsymbol{x}-\boldsymbol{x}'|=\frac{\boldsymbol{I}}{2\pi}\nabla\log|\boldsymbol{x}-\boldsymbol{x}'|\\&=\frac{\boldsymbol{I}\cdot(\boldsymbol{x}-\boldsymbol{x}')}{2\pi|\boldsymbol{x}-\boldsymbol{x}'|^2}.\end{aligned}\tag{3.1.11}$$

This result corresponds to the second term in the expansion (2.1.28) and also to the first-order solution ψ_1 of the Laplace equation. In polar coordinates with origin at the centre of the dipole and the vector $\boldsymbol{I}$ directed along the axis $\theta=\frac{1}{2}\pi$, the stream function ψ and radial u and azimuthal v velocity components are given by

$$\left.\begin{aligned}\psi&=\frac{I}{2\pi r}\sin\theta,\\ \begin{bmatrix}u\\ v\end{bmatrix}&=\frac{I}{2\pi r^2}\begin{bmatrix}\cos\theta\\ \sin\theta\end{bmatrix},\end{aligned}\right\}\tag{3.1.12}$$

where $I=|\boldsymbol{I}|$.

In the same way one can obtain more complex point singularities. If a point dipole of intensity $\boldsymbol{I}$ is placed at the point $\boldsymbol{x}'+\frac{1}{2}\delta\boldsymbol{x}'$ and a dipole of equal strength but opposite direction $-\boldsymbol{I}$ is placed at the point $\boldsymbol{x}'-\frac{1}{2}\delta\boldsymbol{x}'$ then on decreasing $|\delta\boldsymbol{x}'|$ and increasing $|\boldsymbol{I}|$ such that their product remains finite,

$$\lim_{|\delta \boldsymbol{x}'| \to 0, |\boldsymbol{I}| \to \infty} I_i \delta x'_j = M_{ij},$$

a point vortex quadrupole is obtained. The flow field induced by the quadrupole is given by the stream function

$$\psi = M_{ij} \frac{\partial^2}{\partial x_i \partial x_j} \left(-\frac{\log |\boldsymbol{x} - \boldsymbol{x}'|}{2\pi} \right) = M_{ij} \frac{n_i n_j}{4\pi |\boldsymbol{x}|^2} \tag{3.1.13}$$

and corresponds to the third term in (2.1.28). Obviously this singularity can also be considered as a superposition of four point vortices placed at the apexes of a parallelogram near the point $\boldsymbol{x}'$. The distribution of velocity connected with this type of singularity is discussed in section 3.5.3.

3.1.3 The point vortex couple and the vortex dipole with distributed vorticity

Consider two point vortices of equal but oppositely signed strengths Γ and $-\Gamma$ without taking the limit (3.1.11). Let $a = |\delta \boldsymbol{x}'|$ be the distance between the vortices. Then the velocity of each vortex at the location of the other is $\Gamma/2\pi a$ and is directed in the same sense. The entire system therefore translates at a speed $U = \Gamma/2\pi a$ relative to the fluid at infinity. To find the stream function in the frame co-moving with this point vortex couple, one must subtract the velocity U. When the vortices of strengths $-\Gamma$ and Γ are located at $(r, \theta) = (\frac{1}{2}a, \frac{1}{2}\pi)$ and $(\frac{1}{2}a, -\frac{1}{2}\pi)$ respectively, the stream function ψ associated with the flow is given by (Lamb, 1932)

$$\left.\begin{aligned} \psi &= \frac{\Gamma}{2\pi} \left(-\frac{r \sin\theta}{a} + \log \frac{|\boldsymbol{x} + \frac{1}{2}\delta \boldsymbol{x}'|}{|\boldsymbol{x} - \frac{1}{2}\delta \boldsymbol{x}'|} \right) \\ &= \frac{\Gamma}{2\pi} \left(-\frac{r \sin\theta}{a} + \frac{1}{2} \log \frac{\frac{1}{4}a^2 + ar\sin\theta + r^2}{\frac{1}{4}a^2 - ar\sin\theta + r^2} \right). \end{aligned}\right\} \tag{3.1.14}$$

The streamline pattern of this flow is shown in Figure 3.1. Here we have a uniform stream $(-U)$ at infinity. The line $\psi = 0$ consists partly of the axes $\theta = 0, \pi$ and partly of an oval around both of the vortices. It is clear that the closed streamline acts as a solid boundary

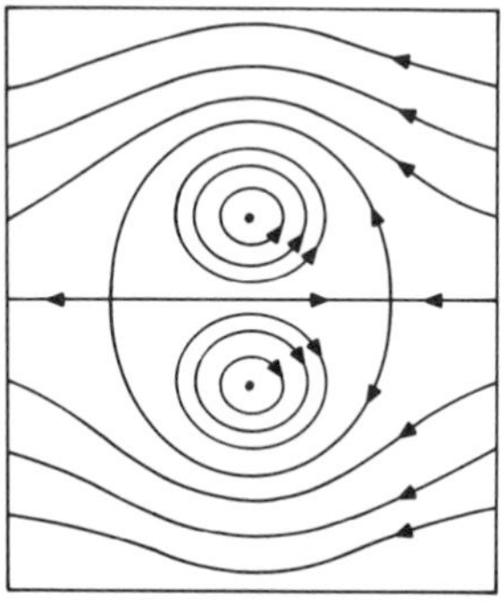

Figure 3.1 *Streamline pattern of a freely moving point vortex couple, according to (3.1.14).*

to the fluid particles inside it, and in the case of a translating couple all interior particles remain trapped within this closed region, which is sometimes referred to as the 'atmosphere' of the couple. The motion outside the oval is exactly the same as the irrotational motion around a rigid oval cylinder. The axes of the oval are approximately equal to $2.09a$ and $1.37a$. The total impulse of the flow due to the translating point vortex couple is a vector directed along the axis $\theta = 0$, of magnitude $\rho I = \rho a \Gamma$. Half of the impulse $\frac{1}{2}\rho I = \rho U S$ (where S is the area of the oval) is connected with the translational motion of the fluid in the oval, while the remaining half is connected with the pressure distribution in the fluid (section 3.4).

The combination of two point vortices of opposite strengths is the simplest model of a steady vortex structure that possesses a momentum. Another example is a steady dipole vortex with a continuous vorticity distribution in a closed region. Assuming a linear relationship

$$\omega = -k^2 \psi \tag{3.1.15}$$

inside the region where the vorticity is located, one finds, after solving the kinematic equation (3.0.3), that this region is a circle of radius $R = 3.83k^{-1}$, where k is some constant and $Rk = c = 3.83...$ is the first zero of the first-order Bessel function J_1. In a polar coordinate system connected with a circle translating with speed U in the direction $\theta = 0$ the stream function of the flow is given by (e.g.

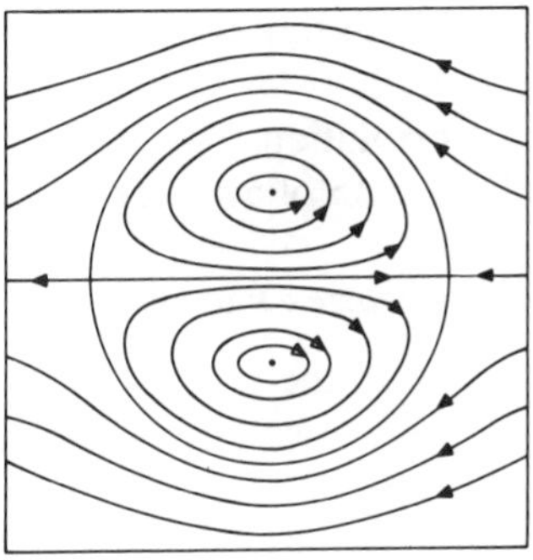

Figure 3.2 *Streamline pattern of a distributed vortex dipole in a uniform irrotational stream, according to (3.1.16).*

Batchelor, 1967)

$$\left.\begin{aligned} \psi_e &= -U\left(r - \frac{R^2}{r}\right)\sin\theta \qquad (r > R), \\ \psi_i &= -\frac{2URJ_1(cr/R)}{cJ_0(c)}\sin\theta \quad (r < R). \end{aligned}\right\} \tag{3.1.16}$$

Here ψ_e describes the external irrotational motion similar to that around a cylinder of radius R translating with speed U, while ψ_i represents the internal vortical dipole motion (Figure 3.2). The values of the stream functions ψ_e and ψ_i and also the azimuthal velocities are required to match smoothly at the circle $r = R$, while the vorticity is discontinuous at this contour. The impulse vector related to the translating dipole is directed along the axis $\theta = 0$ and has magnitude $2\pi R^2 \rho U$.

3.2 Diffusion of a vortex line and other vortex monopoles in a viscous fluid

As mentioned above, molecular diffusion in a real viscous fluid smooths discontinuities of shear stresses and some flow characteristics such as vorticity that may appear at the initial moment of time. Thus a flow in a real fluid becomes essentially unsteady. Also, the additional term on the right-hand side of the equation of motion describing the effects of viscosity increases the order of the differential equation. Both of these circumstances essentially complicate the problem of finding exact solutions for vortex multipoles. However, in some simple cases these solutions can be found. The simplest example is the so-called diffusion of a vortex line.

3.2.1 *The flow induced by a vortex monopole*

Suppose that initially, at $t = 0$, there is a distribution of velocity corresponding to a rectilinear vortex line of intensity Γ. In a plane perpendicular to the line the velocity components at $t = 0$ are determined by (3.1.10). Consider the evolution of the flow with time. Since the initial and boundary conditions do not depend on the polar angle θ, one seeks a solution with the same property. It is also clear that the radial component of velocity must remain zero. For the azimuthal component of velocity the momentum equation (2.1.14) gives

$$\frac{\partial v}{\partial t} = \nu\left(\frac{\partial^2 v}{\partial r^2} + \frac{1}{r}\frac{\partial v}{\partial r} - \frac{1}{r^2}v\right). \tag{3.2.1}$$

To decrease the number of arguments (r, t) from two to one, let us use dimensional analysis. Generally the unknown function v depends on four governing parameters: ν, Γ, t and r. Two of these have independent dimensions, while the dimensions of Γ and ν are the same (L^2T^{-1}). Since v is initially proportional to Γ, we can write, after applying the Π theorem,

$$v(r,t) = \frac{\Gamma}{\nu}\left(\frac{\nu}{t}\right)^{1/2} V(\eta), \qquad \eta = \frac{r}{2(\nu t)^{1/2}}, \tag{3.2.2}$$

where η is a non-dimensional similarity variable and $V(\eta)$ is the unknown non-dimensional function. Substitution of (3.2.2) into (3.2.1) gives for V the ordinary differential equation

$$\frac{d^2V}{d\eta^2} + \left(\frac{1}{\eta} + 2\eta\right)\frac{dV}{d\eta} - \left(\frac{1}{\eta^2} - 2\right)V = 0.$$

On substituting $U(x) = \eta V(\eta)$ and $x = \eta^2$ we obtain a simple equation for U:

$$x^{1/2}\left(\frac{d^2U}{dx^2} + \frac{dU}{dx}\right) = 0,$$

which can easily be integrated. Finally, for $V(\eta)$ we find the general solution in the form

$$V = \frac{1}{\eta}(C_1\,\mathrm{e}^{-\eta^2} + C_2), \tag{3.2.3}$$

where C_1 and C_2 are constants of integration. In setting up the problem, we have implied that the solution for v must not have a singularity at $r = 0$ ($\eta = 0$) for $t > 0$. This gives $C_1 = -C_2$ in (3.2.3). The initial condition (3.1.10) gives $C_2 = 1/4\pi$. Thus we arrive at the azimuthal flow velocity distribution

$$v(r, t) = \frac{\Gamma}{2\pi r}(1 - \mathrm{e}^{-r^2/4\nu t}) \tag{3.2.4}$$

(which is shown graphically in Figure 3.3a) and also the vorticity distribution (Figure 3.3b)

$$\omega = \frac{1}{r}\frac{\partial}{\partial r}(rv) = \frac{\Gamma}{4\pi\nu t}\,\mathrm{e}^{-r^2/4\nu t}. \tag{3.2.5}$$

In practice, one can try to generate such a flow in the following simple way. A thin needle of radius r_0 is rotated with angular velocity Ω_0 in a fluid at rest. Then, on decreasing r_0 and simultaneously

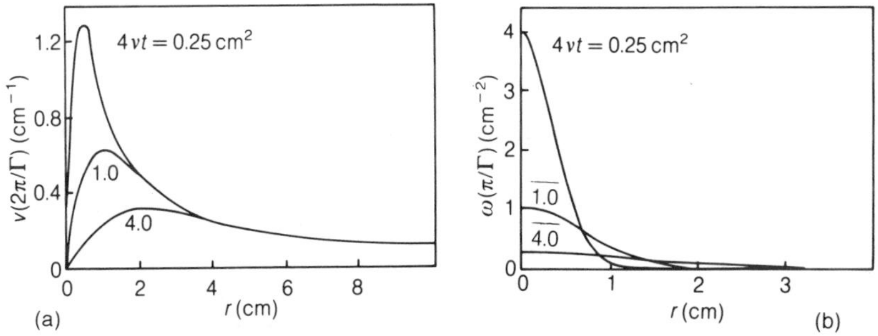

Figure 3.3 *Radial distributions of the azimuthal velocity (a) and vorticity (b) associated with the diffusion of the point vortex (3.2.4) and (3.2.5). The velocity and vorticity have been normalized.*

increasing Ω_0 such that the product $r_0^2\Omega_0$ remains constant,

$$\Gamma = 2\pi \lim_{r_0 \to 0,\ \Omega_0 \to \infty} r_0^2 \Omega_0, \tag{3.2.6}$$

a 'point' source of vorticity with intensity Γ is obtained.

Suppose that the source starts at $t=0$ in a fluid initially at rest. The general solution (3.2.3) must also give the solution of this particular problem. The initial condition $v(r,0)=0$ $(r>0)$ gives $C_2=0$ in (3.2.3). Using the boundary condition (3.2.6), we find $C_1=1/4\pi$. Thus the distributions of the azimuthal velocity and vorticity in a flow induced by the source considered are

$$v = \frac{\Gamma}{2\pi r} e^{-r^2/4\nu t}, \qquad \omega = -\frac{\Gamma}{4\pi \nu t} e^{-r^2/4\nu t}. \tag{3.2.7}$$

The moment of forces $\rho\mathcal{M}$ that must be applied to the needle (per unit length) to supply its rotation is equal in magnitude and opposite in sign to the moment of frictional forces acting on the needle as it rotates in the viscous fluid:

$$\rho\mathcal{M} = -\lim_{r=r_0 \to 0} \left[2\pi\nu r^2 \rho \left(\frac{\partial v}{\partial r} - \frac{v}{r} \right) \right] = 2\nu\rho\Gamma. \tag{3.2.8}$$

The applied moment causes an increase in the angular momentum of the fluid, which was initially at rest. With time, all the vorticity generated by the source is transported away from the latter to infinity by diffusion, and irrotational flow is established asymptotically. This steady flow has a velocity distribution given by

$$v(r,\infty) = \frac{\Gamma}{2\pi r}.$$

If the rotation of the needle is now stopped, the problem becomes that of the diffusion of a vortex line (considered above). Indeed, the boundary condition, now

$$\lim_{r=r_0 \to 0} v = 0 \quad (t>0),$$

gives $C_1 = -C_2$ in (3.2.3), and the initial condition

$$v(r,0) = \frac{\Gamma}{2\pi r}$$

gives $C_2 = 1/4\pi$. Thus the distribution of the azimuthal velocity turns out to be the same as the distribution (3.2.4) for the vortex line.

Using (3.2.8), with the velocity given by (3.2.4), one can easily calculate the moment of frictional forces applied to the needle. It turns out to be zero. Thus, on ceasing to apply the moment $\rho\mathcal{M} = 2\nu\rho\Gamma$ to the needle, it stops immediately and remains at rest. Hence one can easily impart an angular momentum (positive or negative) to the fluid by means of the source considered, but it cannot be extracted passively.

3.2.2 *Motion of a passive tracer in the flow*

To analyse the motion of fluid particles in a flow and to compare indirectly the theoretical solutions with real flows, it is useful to use the so-called Lagrangian flow description. In this one follows the history of individual fluid particles that are 'marked' in some way. Let us mark (e.g. by dye) at some reference time $t = t_0$ particles on a line determined by some relation $\theta_0 = f(r_0)$.

The subsequent motion of each particle is described by the equations of motion

$$\frac{\mathrm{d}r}{\mathrm{d}t} = u(r, \theta, t), \qquad r\frac{\mathrm{d}\theta}{\mathrm{d}t} = v(r, \theta, t) \tag{3.2.9}$$

with initial conditions

$$\theta(t_0) = \theta_0, \qquad r(t_0) = r_0.$$

Here u and v are the radial and azimuthal components of the velocity field, which is assumed to be known. Generally, the system (3.2.9) must be solved numerically. However, in some simple cases an analytical solution can be found.

Consider for example the flow where $u = 0$ and v is given by (3.2.7). Then the first equation (3.2.9) gives

$$r(t) = \text{const} = r_0.$$

Integrating the second equation over time with v given by (3.2.7), we obtain

$$\theta = \theta_0 + \frac{\Gamma}{2\pi r_0^2}\int_{t_0}^{t} \mathrm{e}^{-r_0^2/4\nu t}\,\mathrm{d}t. \tag{3.2.10}$$

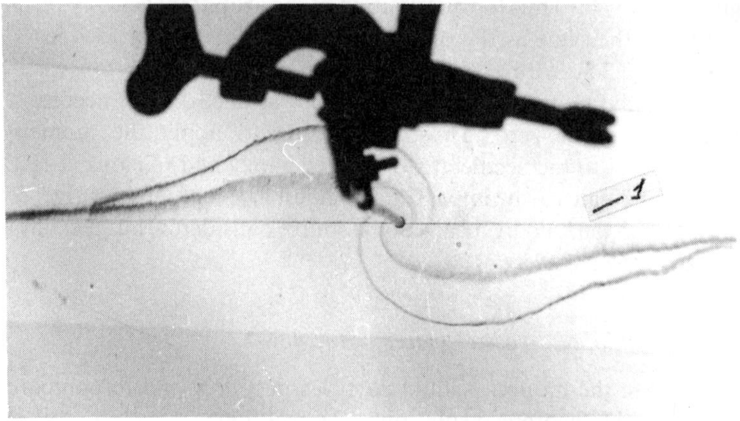

Figure 3.4 *Formation of typical spirals from the straight dyed lines generated periodically across the flow induced by a vortex monopole. Dyed lines are generated using a thymol blue technique (section 1.2.4) by appling a voltage to a thin platinum wire (1).*

Introducing the similarity variable $x = r_0^2/4\nu t$, and denoting $x_0 = r_0^2/4\nu t_0$, we rewrite (3.2.10) in the form

$$\theta = \theta_0 + \frac{\Gamma}{8\pi\nu}\left(\mathrm{e}^{-x}x^{-1} - \mathrm{e}^{-x_0}x_0^{-1} + \int_{x_0}^{x} \mathrm{e}^{-y}y^{-1}\,\mathrm{d}y\right).$$

Thus each marked particle in the flow moves along a circle of radius r_0, rotating at an angle determined by (3.2.10). It is clear that the line of marked particles (e.g. $\theta_0 = 0$, π for $t = t_0$) is transformed with time into a typical spiral (Figure 3.4). That is why flows of this type are sometimes referred to as spiral flows.

3.3 A spiral vortex in a rotating fluid

The family of monopolar vortex flows can be diversified. Consider for example a thin rotating tube with porous walls through which a mass flux of fluid ρq (per unit length of the tube) is injected or withdrawn. Let us decrease the radius r_0 of the tube and increase the angular velocity Ω_0 of rotation and the pressure difference in the tube such that the quantities $\Gamma = 2\pi r_0^2 \Omega_0$ and q remain constant in the limit as $r_0 \to 0$. We thus obtain asymptotically a combined line source of vorticity and mass. In contrast to the cases considered earlier, the radial component of velocity in the flow induced by this source is not zero, and hence the vorticity generated by the source will be transported not only by viscous diffusion but also by advection. This complicates the problem, but not essentially, and the general self-similar solution can be found.

It appears (Afanasyev and Voropayev, 1991) that a more complicated monopolar vortex flow can also be analysed analytically, namely the flow induced by a source of mass in a uniformly rotating fluid. Since in a rotating fluid a source of mass also produces vorticity, this problem includes the previous one. An exact unsteady nonlinear solution for a viscous flow induced in a rotating fluid is rather exceptional in fluid dynamics, and it is worth considering it in some detail.

Thus let us consider a planar flow of a viscous incompressible fluid uniformly rotating with angular velocity $\boldsymbol{\Omega}$. The flow is induced by a line source or sink of mass whose intensity is ρq, the dimensions of q being L^2T^{-1}. The source starts at $t = 0$ and thereafter acts with constant intensity uniformly in all directions. The source is at the point $r = 0$ in a polar coordinate system (r, θ) rotating with the fluid. The fluid is at rest initially and is at rest at infinity.

Since the initial and boundary conditions are independent of θ, let us try to find a solution with the same property. Hence two components of the equation of motion (2.1.16) together with the external body forces and the continuity equation (2.1.8) can be written in the form

$$\frac{\partial u}{\partial t} + u\frac{\partial u}{\partial r} - \frac{1}{r}v^2 - 2\Omega v = -\frac{1}{\rho}\frac{\partial P}{\partial r} + \nu\frac{\partial}{\partial r}\frac{1}{r}\left(\frac{\partial}{\partial r}ru\right), \tag{3.3.1}$$

$$\frac{\partial v}{\partial t} + u\frac{\partial v}{\partial r} + \frac{1}{r}uv + 2\Omega u = \nu\frac{\partial}{\partial r}\frac{1}{r}\left(\frac{\partial}{\partial r}rv\right), \tag{3.3.2}$$

$$\frac{1}{r}\frac{\partial}{\partial r}(ru) = 0; \tag{3.3.3}$$

where u and v are the radial and azimuthal velocity components respectively, and $P = p - \frac{1}{2}\Omega^2 r^2 \rho$ is a reduced pressure. Note that additional terms accounting for the non-inertial rotating coordinate system are included in the momentum equations (3.3.1) and (3.3.2). These terms represent the Coriolis force and the centrifugal force, which is included in the reduced pressure (section 2.1.5).

Integrating the mass conservation equation (3.3.3) over the area of a circle of arbitrary radius centred at $r = 0$ and then transforming the surface integral into a contour integral, we find for the mass flux through the contour a value $2\pi\rho r u$. Comparison of this mass flux with the mass flux ρq supplied by the source gives

$$u = \frac{q}{2\pi r}.$$

Thus the radial component of the velocity is determined, and the nonlinear equation (3.3.2) is essentially simplified:

$$\frac{\partial}{\partial t} v = -\frac{q}{2\pi r^2}\frac{\partial}{\partial r}(rv + r^2\Omega) + \nu\frac{\partial}{\partial r}\frac{1}{r}\left(\frac{\partial}{\partial r} rv\right). \tag{3.3.4}$$

The problem is now to obtain v from (3.3.4) for appropriate initial and boundary conditions. The equation (3.3.1) can then be used for the determination of the pressure distribution.

3.3.1 General self-similar solution

Generally, the unknown azimuthal velocity v depends on five governing parameters: ν, q, Ω, r and t. Two of these have independent dimensions. With the aid of the Π theorem, the unknown function can be represented in the non-dimensional form

$$\frac{v}{(\nu\Omega)^{1/2}} = \Phi(T, \eta, A),$$

where Φ is a non-dimensional function of the non-dimensional arguments T, η and A. Two of these arguments $T = \Omega t$ and $\eta = \frac{1}{2}r(\nu t)^{-1/2}$, are independent variables, while $A = q/2\pi\nu$ is a constant characterizing the intensity of the source. This constant can be considered as an analogue of the Reynolds number of the flow.

In spite of the essential simplification, the problem still remains rather complex. To solve it, let us assume that the variables T and η can be separated, i.e. $\Phi(T, \eta, A) = F(T)V(\eta, A)$, and $F(T)$ can be represented in the form $F(T) = T^{\alpha}$, where α is some unknown constant. If we manage to find a solution of this form that satisfies the initial and boundary conditions, we will have justified this assumption.

Writing v in the form

$$v = (\nu\Omega)^{1/2} T^{\alpha} V(\eta, A) \tag{3.3.5}$$

and substituting this into (3.3.4), we obtain the ordinary differential equation

$$\eta \frac{d^2 V}{d\eta^2} + (1 - A + 2\eta^2)\frac{dV}{d\eta} - \eta\left(\frac{1+A}{\eta^2} + 4\alpha\right)V = 4AT^{1/2-\alpha}.$$

The left-hand side depends only on η, while the right-hand side depends only on T; therefore α must be equal to $\frac{1}{2}$. Using the substitutions $x = \eta^2$ and $U(x) = \eta V$, we rewrite this equation in the form

$$x\frac{d^2 U}{dx^2} + (x - \tfrac{1}{2}A)\frac{dU}{dx} - U = A. \tag{3.3.6}$$

The general solution of (3.3.6) can be expressed in terms of confluent hypergeometric functions. However, there is a simple way to obtain a solution of this equation. Differentiating (3.3.6), we obtain a simple third-order equation

$$\frac{d^3 U}{dx^3} = \frac{d^2 U}{dx^2}\left(\frac{A-2}{2x} - 1\right),$$

whose general solution is given by

$$U = c\int^x \int^y e^{-z} z^{A/2-1}\, dz\, dy + c_1 x + c_2,$$

where c, c_1 and c_2 are constants of integration. Substitution of this solution into (3.3.6) gives

$$c_2 = -A(\tfrac{1}{2}c_1 - 1),$$

and hence

$$U = c\int^x \int^y e^{-z} z^{A/2-1}\, dz\, dy + c_1(x - \tfrac{1}{2}A) - A. \tag{3.3.7}$$

To find the constants c and c_1, one must set appropriate initial and boundary conditions. Although (3.3.7) is the general solution of the equation of motion (3.3.4), it is a self-similar solution and satisfies only those conditions that allow the azimuthal velocity to be presented in the form (3.3.5). Thus, in solving the problem, we have limited the class of possible solutions. The general solution of (3.3.4) must satisfy three independent conditions, e.g.

$$v(r,0)=v_0(r), \qquad v(0,t)=v_1(t), \qquad v(\infty,t)=v_2(t),$$

where v_0, v_1 and v_2 are given functions, while the solution (3.3.7) need satisfy only two conditions, e.g.

$$U(0)=U_0, \qquad U(\infty)=U_1.$$

This is clearly the price paid for the simplifications that were made in order to obtain a solution relatively easily. However, (3.3.7) is still quite general. In particular, it describes a wide class of motions in which the fluid is initially at rest and remains at rest at infinity, i.e.

$$\begin{aligned} v(r,0)&=0 \quad (r\neq 0),\\ v(\infty,t)&=0 \quad (t>0). \end{aligned}$$

In the new variables these two conditions can be expressed as the single condition

$$x^{-1/2}U=0 \quad (x\to\infty), \qquad x=\frac{r^2}{4\nu t},$$

which gives $c_1=0$ in (3.3.7), so that U can be expressed in a more convenient form as

$$U=c\int_\infty^x \mathrm{e}^{-y}y^{A/2-1}(x-y)\,\mathrm{d}y-A. \tag{3.3.8}$$

To find the remaining constant c, one must set an appropriate condition at $x=0$. The velocity v and vorticity ω can then be obtained from the expressions

$$\left.\begin{aligned} v&=(\nu\Omega)^{1/2}T^{1/2}x^{-1/2}U(x,A),\\ \omega&=\frac{1}{r}\frac{\partial}{\partial r}(rv)=\Omega\frac{\mathrm{d}U}{\mathrm{d}x}. \end{aligned}\right\} \tag{3.3.9}$$

3.3.2 *Some particular cases*

Consider two examples. Let the source of mass be a thin porous tube that can rotate freely on ideal bearings. It is clear that such a source does not apply any moment to the ambient fluid, so that

$$-2\pi\rho\nu r^2\left(\frac{\partial v}{\partial r}-\frac{v}{r}\right)=0 \quad (r\to 0).$$

Rewriting this condition in the new variables

$$U-x\frac{\mathrm{d}U}{\mathrm{d}x}=0 \quad (x\to 0),$$

we can find the constant c in (3.3.8) and obtain the solution of this

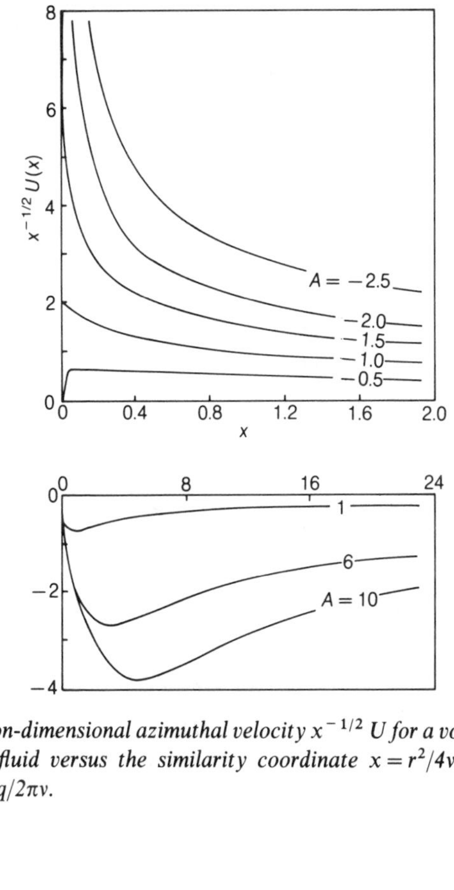

Figure 3.5 *Non-dimensional azimuthal velocity* $x^{-1/2}U$ *for a vortex monopole in a rotating fluid versus the similarity coordinate* $x=r^2/4\nu t$, *for different values of* $A=q/2\pi\nu$.

problem in the form

$$\left.\begin{aligned} U &= A\left[\left(\int_{\infty}^{x} \mathrm{e}^{-y}\, y^{(A-2)/2}(y-x)\,\mathrm{d}y\right)\left(\int_{\infty}^{0} \mathrm{e}^{-y}\, y^{A/2}\,\mathrm{d}y\right)^{-1} - 1\right],\\ \frac{\omega}{\Omega} &= \frac{\mathrm{d}U}{\mathrm{d}x} = -A\left(\int_{\infty}^{x} \mathrm{e}^{-y}\, y^{(A-2)/2}\,\mathrm{d}y\right)\left(\int_{\infty}^{0} \mathrm{e}^{-y}\, y^{A/2}\,\mathrm{d}y\right)^{-1}. \end{aligned}\right\} \tag{3.3.10}$$

Graphs of the non-dimensional azimuthal velocity $v/\Omega(\nu t)^{1/2} = x^{-1/2}U$ for different A are shown in Figure 3.5. Similar distributions of non-dimensional vorticity $\omega/\Omega = \mathrm{d}U/\mathrm{d}x$ are shown in Figure 3.6. The source of mass in the rotating fluid generates

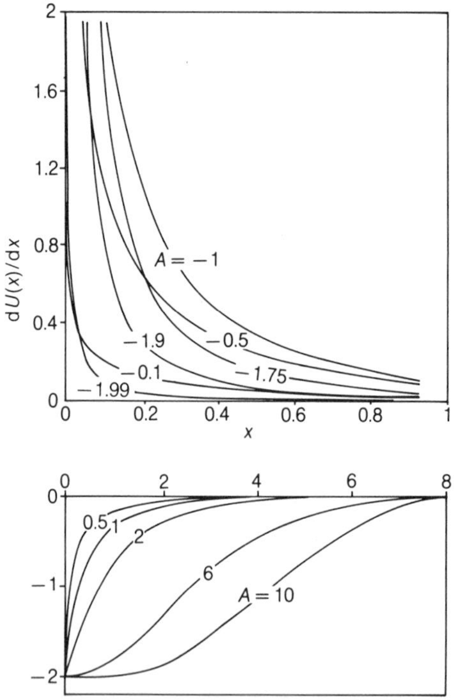

Figure 3.6 *Non-dimensional vorticity* $\mathrm{d}U/\mathrm{d}x$ *for a vortex monopole in a rotating fluid versus the similarity coordinate* $x = r^2/4\nu t$ *for different values of* $A = q/2\pi\nu$.

vorticity, which is then transported by viscous diffusion and advection. When $A > 0$ (source of mass) both effects have the same sign because the radial velocity is directed from the source. In this case the vorticity in the vicinity of the source is constant and does not depend on $A: \mathrm{d}U/\mathrm{d}x = -2$ for $x = 0$. When $A > 2$, advective transport prevails and the vorticity progresses like a nonlinear wave with a steep front. The non-dimensional position $x_* = r_*^2/4\nu t$ of the wave front, where $\mathrm{d}^3U/\mathrm{d}x^3 = 0$, is $x_* = \frac{1}{2}A - 1$, which gives a propagation velocity for the wave of

$$u_* = \frac{\mathrm{d}r_*}{\mathrm{d}t} = \left[(\tfrac{1}{2}A - 1)\frac{\nu}{t} \right]^{1/2}.$$

When $0 < A < 2$, viscous diffusion prevails and the wave does not form.

When $A < 0$ (sink of mass) the radial component of velocity is directed towards the origin, and the effects of viscous diffusion and advection are in opposition. The vorticity at the origin becomes infinitely large:

$$\frac{\mathrm{d}U}{\mathrm{d}x} \sim \frac{-2x^{A/2}}{\displaystyle\int_\infty^0 \mathrm{e}^{-y} y^{A/2}\,\mathrm{d}y} \to \infty \quad (x \to 0).$$

When $A = -2$, these effects compensate one another. When $A < -2$, all the vorticity is swept by the flow into the sink. In this case the vorticity is zero everywhere except at the point $x = 0$, where a 'non-diffusive' vortex line with increasing intensity $\Gamma = -4\pi A \nu \Omega t$ forms.

Now consider the case where the source of mass in the form of a thin porous tube is fixed and cannot rotate. The no-slip condition at the surface of the tube gives the boundary condition in the form

$$U = 0 \quad (x = 0).$$

Using this condition, we can find the constant c in (3.3.8) and obtain the solution of the problem in the form

$$U = A\left[\left(\int_\infty^x \mathrm{e}^{-y} y^{(A-2)/2}(x-y)\,\mathrm{d}y\right)\left(\int_\infty^0 \int_\infty^y \mathrm{e}^{-z} z^{(A-2)/2}\,\mathrm{d}z\,\mathrm{d}y\right)^{-1} - 1\right],$$

$$\frac{dU}{dx} = A\left(\int_{\infty}^{x} e^{-y} y^{(A-2)/2}\, dy\right)\left(\int_{\infty}^{0}\int_{\infty}^{y} e^{-z} z^{(A-2)/2}\, dz\, dy\right)^{-1}. \tag{3.3.11}$$

Note that the expressions (3.3.10) and (3.3.11) are equal for all x except $x=0$ when $A<-2$. At this point the integrals in the denominators in (3.3.10) and (3.3.11) have different asymptotic behaviour. Comparing these denominators, we find

$$\lim_{x\to 0} \frac{\displaystyle\int_{\infty}^{x}\int_{\infty}^{y} e^{-z} z^{(A-2)/2}\, dz\, dy}{\displaystyle\int_{\infty}^{x} e^{-y} y^{A/2}\, dy} = \begin{cases} -1 & (A \geqslant -2), \\ 2/A & (A < -2). \end{cases}$$

Using the solutions obtained, it is possible to consider the balance of angular momentum in the system. For example, let the fluid be sucked ($A<0$) into a sink in the form of a thin porous tube, which is fixed. The flux of absolute angular momentum swept with the fluid into the sink is given by

$$-q\rho(rv + r^2\Omega) = -4\pi\rho v^2 \Omega A t(U+2x) \quad (x \to 0),$$

where U is determined by (3.3.11). Using the fact that $U \to 0$ as $x \to 0$, we find that this flux tends to zero as $x \to 0$. It appears that all angular momentum transported with the sucking fluid to the sink is absorbed by the surface of the fixed tube because of the no-slip condition here. With the help of (3.2.8), the moment of the frictional forces applied per unit length of the sink is found to be

$$2\pi\rho v \lim_{r\to 0} r^2\left(\frac{\partial v}{\partial r} - \frac{v}{r}\right) = 8\pi\rho v^2 \Omega t \lim_{x\to 0}\left(x\frac{dU}{dx} - U\right)$$

$$= \begin{cases} 0 & (-2 \leqslant A < 0) \\ 4\pi\rho v^2 \Omega t(A+2)A & (A < -2). \end{cases} \tag{3.3.12}$$

Thus when the sink is sufficiently strong ($A<-2$, $q<-4\pi v$), the moment applied to the fixed tube increases linearly with time.

In contrast, the flux of absolute angular momentum that is swept with the fluid into the tube, which rotates freely, is equal to the moment given by (3.3.12), as one would expect.

3.4 Impulse integral and flow momentum

In section 2.1.7 the multipole expansion (2.1.28) was derived from just the kinematic relation between the stream function and vorticity in the form of the Poisson equation (2.1.25). However, one must use the dynamical equations to clarify the meaning of the numerators in the expansion.

Consider the planar unsteady flow of a viscous incompressible fluid induced by a system of forces $\rho \boldsymbol{f}(\boldsymbol{x}', t)$ (force per unit area of unit thickness) that acts on the fluid inside the area $S'(\boldsymbol{x}')$ (Figure 2.1). The forces and associated vorticity distribution $\boldsymbol{\omega}(\boldsymbol{x}, t) = (\boldsymbol{k}\omega)$ are assumed to occupy a finite area $S'(\boldsymbol{x}')$ and the fluid extends to infinity, $\boldsymbol{k}$ being the unit vector along the x_3 axis perpendicular to the plane of motion.

The numerator Γ of the first term in (2.1.28) is rather simple: it is an integral of the vorticity distribution in the flow. Let us convince ourselves first that $\Gamma = \text{const}$ for the considered flows. Using the definition of $\boldsymbol{\omega}$ and Stokes' theorem, we have

$$\frac{\mathrm{d}}{\mathrm{d}t}\int_S \boldsymbol{\omega}\,\mathrm{d}S = \int_S (\boldsymbol{\nabla}\times\frac{\partial}{\partial t}\boldsymbol{u})\,\mathrm{d}S = \int_L \left(\boldsymbol{n}\times\frac{\partial}{\partial t}\boldsymbol{u}\right)\mathrm{d}L,$$

where $\boldsymbol{n} = \boldsymbol{x}/|\boldsymbol{x}|$, $\mathrm{d}L = R\,\mathrm{d}\theta$ and L is a circle of radius R. For large distances $|\boldsymbol{x}| = R$ the first term in the expansion (2.1.28) gives $\boldsymbol{u} \sim 1/R$ and the momentum equation (2.1.16) becomes

$$\frac{\partial \boldsymbol{u}}{\partial t} = -\frac{1}{\rho}\boldsymbol{\nabla}p.$$

Hence for $R \to \infty$

$$\frac{\mathrm{d}}{\mathrm{d}t}\int_S \boldsymbol{\omega}\mathrm{d}S = -\frac{1}{\rho}\int_L (\boldsymbol{n}\times\boldsymbol{\nabla}p)\,\mathrm{d}L = -\frac{\boldsymbol{k}}{\rho}\int_0^{2\pi}\frac{\partial p}{\partial\theta}\mathrm{d}\theta = 0.$$

Thus the considered system of forces cannot change the total vorticity of the flow.

It turns out that the numerator (2.1.30) of the second term is determined by the impulse of external forces applied on the fluid. Following Cantwell (1986), we consider this in more detail in terms of the vector potential (3.0.1) introduced above. The vector potential $\boldsymbol{B}(\boldsymbol{x}, t) = (0, 0, \psi) = \boldsymbol{k}\psi$, where $\psi(\boldsymbol{x}, t)$ is the stream function at a point

$\boldsymbol{x}$, can be found from the vector Poisson equation

$$\nabla^2 \boldsymbol{B} = -\boldsymbol{\omega},$$

whose solution for two dimensions is

$$\boldsymbol{B}(\boldsymbol{x}, t) = -\frac{1}{2\pi}\int_{S'} \boldsymbol{\omega}(\boldsymbol{x}', t)\log|\boldsymbol{x} - \boldsymbol{x}'|\,\mathrm{d}S'. \tag{3.4.1}$$

We want to find an expression for the volume-integrated (per unit thickness of the fluid) momentum divided by the density given by

$$\boldsymbol{H}(t) = \int_S \boldsymbol{u}(\boldsymbol{x}, t)\,\mathrm{d}S. \tag{3.4.2}$$

Taking into account that, by definition,

$$\boldsymbol{u} = \nabla \times \boldsymbol{B},$$

the surface integral (3.4.2) can be represented as a contour integral of the vector potential:

$$\boldsymbol{H}(t) = \int_L \boldsymbol{n} \times \boldsymbol{B}\,\mathrm{d}L, \tag{3.4.3}$$

where $\boldsymbol{n} = \boldsymbol{x}/|\boldsymbol{x}| = \boldsymbol{i}\cos\theta + \boldsymbol{j}\sin\theta$, $\boldsymbol{i}$ and $\boldsymbol{j}$ being unit vectors, $|\boldsymbol{x}| = R$, $\mathrm{d}L = R\,\mathrm{d}\theta$ and L is a circle of radius R. Substituting the expression for $\boldsymbol{B}$ from (3.4.1) into (3.4.3), exchanging the order of integration and using the fact that

$$\int_L \boldsymbol{x}\log|\boldsymbol{x} - \boldsymbol{x}'|\,\mathrm{d}\theta$$

$$= \tfrac{1}{2}R\int_0^{2\pi} (\boldsymbol{i}\cos\theta + \boldsymbol{j}\sin\theta)\left[\log R^2 + \log\left|1 + \left(\frac{r'}{R}\right)^2 - 2\frac{r'}{R}\cos(\theta - \theta')\right|\right]\mathrm{d}\theta$$

$$= \tfrac{1}{2}R\int_0^{2\pi} [\boldsymbol{i}\cos(\theta + \theta') + \boldsymbol{j}\sin(\theta + \theta')]\log\left|1 + \left(\frac{r'}{R}\right)^2 - 2\frac{r'}{R}\cos\theta\right|\mathrm{d}\theta$$

$$= \tfrac{1}{2}R \int_0^{2\pi} [\boldsymbol{i}(\cos\theta\cos\theta' - \sin\theta\sin\theta')$$

$$+ \boldsymbol{j}(\sin\theta\cos\theta' + \cos\theta\sin\theta')] \log\left|1 + \left(\frac{r'}{R}\right)^2 - 2\frac{r'}{R}\cos\theta\right| \mathrm{d}\theta$$

$$= \tfrac{1}{2}R(\boldsymbol{i}\cos\theta' + \boldsymbol{j}\sin\theta') \int_0^{2\pi} \cos\theta \log\left|1 + \left(\frac{r'}{R}\right)^2 - 2\frac{r'}{R}\cos\theta\right| \mathrm{d}\theta$$

$$= -\pi r'(\boldsymbol{i}\cos\theta' + \boldsymbol{j}\sin\theta') = -\pi \boldsymbol{x}'$$

(for the last integral see e.g. Dwight, 1961) leads to

$$\int_L \boldsymbol{n} \times \boldsymbol{B}\,\mathrm{d}L = -\frac{1}{2\pi}\int_{S'}\int_L \boldsymbol{x} \times \boldsymbol{\omega}(\boldsymbol{x}', t) \log|\boldsymbol{x} - \boldsymbol{x}'|\,\mathrm{d}S'\,\mathrm{d}\theta$$

$$= \tfrac{1}{2}\int_{S'} \boldsymbol{x}' \times \boldsymbol{\omega}(\boldsymbol{x}', t)\,\mathrm{d}S'.$$

Thus

$$\boldsymbol{H}(t) = \tfrac{1}{2}\boldsymbol{I}(t), \tag{3.4.4}$$

where

$$\boldsymbol{I}(t) = \int_S \boldsymbol{x} \times \boldsymbol{\omega}(\boldsymbol{x}, t)\,\mathrm{d}S \tag{3.4.5}$$

is called the impulse of the vorticity distribution (Lamb, 1932). The integral of the momentum is fully convergent as long as the circle contains all of the vorticity. Irrotational flow motions beyond the vortical region do not contribute to the total momentum.

At large values of $|\boldsymbol{x}|$ the vector potential may be approximated, similarly to the expansion (2.1.28), by the first few terms of a multipole expansion. The facts that $\boldsymbol{\omega}$ is divergence-free, $\nabla\cdot\boldsymbol{\omega} = \nabla\cdot(\nabla\times\boldsymbol{u}) = 0$, and localized and that $\boldsymbol{f}(\boldsymbol{x}', t)$ produces no net vorticity in the fluid imply that the first term of this expansion vanishes: $\boldsymbol{\Gamma} = \int_{S'}\boldsymbol{\omega}(\boldsymbol{x}', t)\,\mathrm{d}S' = 0$. Using a Taylor series similar to (2.1.27) and the fact that

$$\frac{\boldsymbol{x}}{|\boldsymbol{x}|} \times (\boldsymbol{x}' \times \boldsymbol{\omega}) = \boldsymbol{n} \times (\boldsymbol{x}' \times \boldsymbol{\omega}) = -(\boldsymbol{n}\cdot\boldsymbol{x}')\boldsymbol{\omega},$$

the second term in the expansion can be represented in the form

$$\frac{1}{2\pi}\frac{\partial \log|\boldsymbol{x}|}{\partial \boldsymbol{x}_i}\int_{S'} x_i'\boldsymbol{\omega}(\boldsymbol{x}',t)\,\mathrm{d}S' = \frac{n_i}{2\pi|\boldsymbol{x}|}\int_{S'} x_i'\boldsymbol{\omega}(\boldsymbol{x}',t)\,\mathrm{d}S'$$

$$= \frac{1}{2\pi|\boldsymbol{x}|}\int_{S'} (\boldsymbol{n}\cdot\boldsymbol{x}')\,\boldsymbol{\omega}(\boldsymbol{x}',t)\,\mathrm{d}S'$$

$$= \frac{\boldsymbol{I}\times\boldsymbol{n}}{2\pi|\boldsymbol{x}|}.$$

Thus at large $|\boldsymbol{x}|$ the vector potential is, to lowest order,

$$\boldsymbol{B} = \frac{\boldsymbol{I}\times\boldsymbol{n}}{2\pi|\boldsymbol{x}|} + O\left(\frac{1}{|\boldsymbol{x}|^2}\right), \tag{3.4.6}$$

where $\boldsymbol{I}$ is defined by (3.4.5).

In order to evaluate the impulse integral (3.4.5) via the external forcing, we need to perform an integral momentum balance over S. In two dimensions the condition of momentum balance over S can be written in a form similar to (2.1.17):

$$\frac{\mathrm{d}}{\mathrm{d}t}H_i(t) + \int_L \frac{1}{\rho}\Pi_{ik}n_k\,\mathrm{d}L = \int_S f_i\,\mathrm{d}S.$$

Using (2.1.9)–(2.1.11), this condition takes the form

$$\frac{\mathrm{d}}{\mathrm{d}t}H_i(t) + \int_L \left[u_iu_j + \frac{1}{\rho}\delta_{ij}p - \nu\left(\frac{\partial u_i}{\partial x_j} + \frac{\partial u_j}{\partial x_i}\right)\right]n_j\,\mathrm{d}L = \int_S f_i\,\mathrm{d}S. \tag{3.4.7}$$

For large $|\boldsymbol{x}| = R$ (3.4.6) gives $\boldsymbol{u} \sim 1/R^2$, and we can estimate the boundary integrals of the nonlinear and viscous terms in (3.4.7) as

$$\lim_{R\to\infty}\int_L u_iu_jn_j\,\mathrm{d}L \sim \frac{1}{R^3},$$

$$\lim_{R\to\infty}\int_L \left(\frac{\partial u_i}{\partial x_j} + \frac{\partial u_j}{\partial x_i}\right)n_j\,\mathrm{d}L \sim \frac{1}{R^2},$$

Thus, as $R\to\infty$,

$$\frac{\mathrm{d}}{\mathrm{d}t}\boldsymbol{H} + \int_L \left(\frac{p}{\rho}\right)\boldsymbol{n}\,\mathrm{d}L = \int_S \boldsymbol{f}(\boldsymbol{x},t)\,\mathrm{d}S. \tag{3.4.8}$$

At large $|\boldsymbol{x}|$ the momentum equation is

$$\frac{\partial \boldsymbol{u}}{\partial t} + \frac{1}{\rho} \boldsymbol{\nabla} p = 0 \tag{3.4.9}$$

and we can use (3.4.9) to determine an expression for the far-field pressure. Using (3.4.6) for the vector potential and the fact that

$$\boldsymbol{\nabla} \frac{\boldsymbol{x}}{|\boldsymbol{x}|^2} = \boldsymbol{\nabla} \times \left(\frac{\boldsymbol{x}}{|\boldsymbol{x}|^2} \right) = 0$$

for $\boldsymbol{x} = (x_1, x_2, 0)$, we can write the velocity $\boldsymbol{u} = \boldsymbol{\nabla} \times \boldsymbol{B}$ as the gradient of a scalar. At large $|\boldsymbol{x}|$

$$\begin{aligned} \boldsymbol{u} &= \frac{1}{2\pi} \left[\boldsymbol{I} \left(\boldsymbol{\nabla} \cdot \frac{\boldsymbol{x}}{|\boldsymbol{x}|^2} \right) - (\boldsymbol{I} \cdot \boldsymbol{\nabla}) \frac{\boldsymbol{x}}{|\boldsymbol{x}|^2} \right] \\ &= -\frac{1}{2\pi} \left[(\boldsymbol{I} \cdot \boldsymbol{\nabla}) \frac{\boldsymbol{x}}{|\boldsymbol{x}|^2} + \boldsymbol{I} \times \left(\boldsymbol{\nabla} \times \frac{\boldsymbol{x}}{|\boldsymbol{x}|^2} \right) \right] \\ &= -\frac{1}{2\pi} \boldsymbol{\nabla} \left(\frac{\boldsymbol{I} \cdot \boldsymbol{x}}{|\boldsymbol{x}|^2} \right). \end{aligned} \tag{3.4.10}$$

Substituting (3.4.10) into (3.4.9) and solving, we have, to within an additive function of time,

$$\frac{p}{\rho} = \frac{1}{2\pi} \frac{\mathrm{d}\boldsymbol{I}}{\mathrm{d}t} \cdot \frac{\boldsymbol{x}}{|\boldsymbol{x}|^2}.$$

The boundary integral of the pressure in (3.4.8) may now be evaluated as

$$\begin{aligned} \int_L \frac{p}{\rho} \boldsymbol{n} \, \mathrm{d}L &= \frac{1}{2\pi} \int_L \left(\frac{\mathrm{d}\boldsymbol{I}}{\mathrm{d}t} \cdot \boldsymbol{x} \right) \frac{\boldsymbol{x}}{|\boldsymbol{x}|^3} \, \mathrm{d}L \\ &= \frac{1}{2\pi} \frac{\mathrm{d}|\boldsymbol{I}|}{\mathrm{d}t} \int_0^{2\pi} (\cos\theta_0 \cos\theta + \sin\theta_0 \sin\theta)(\boldsymbol{i} \cos\theta + \boldsymbol{j} \sin\theta) \, \mathrm{d}\theta \\ &= \frac{1}{2} \frac{\mathrm{d}\boldsymbol{I}}{\mathrm{d}t}, \end{aligned} \tag{3.4.11}$$

where the integral is over the circle of radius $R \to \infty$ and $\boldsymbol{I} = |\boldsymbol{I}| (\boldsymbol{i} \cos\theta_0 + \boldsymbol{j} \sin\theta_0)$. Substituting (3.4.4) and (3.4.11) into (3.4.8) and integrating over time, we obtain

$$\boldsymbol{I}(t) = \int_0^t \int_S \boldsymbol{f}(\boldsymbol{x}, t)\, \mathrm{d}S\, \mathrm{d}t. \tag{3.4.12}$$

Thus the integral $\boldsymbol{I}(t)$ is the total impulse applied by the force distribution since the onset of the motion. According to (3.4.4) and (3.4.12), half of the applied impulse ends up in the momentum of the fluid; the remaining half is removed by the pressure field at infinity, which opposes the motion.

3.5 Stokes multipoles

It contrast to the problem of evolution of vortex monopoles, considered in section 3.2, a similar consideration of higher-order multipoles presents definite difficulties. Since a point vortex induces only azimuthal motion, and the radial component of velocity is either zero or can be determined independently as in the case of a spiral vortex in a rotating fluid, the nonlinear terms in the momentum equation vanish or can be represented in linearized form. Thus the problem of evolution of a vortex monopole is a linear one. The higher-order multipoles induce both components of velocity, and the problem is nonlinear. The full nonlinear unsteady equation is too complicated for analysis. Nevertheless, it is useful to make a first step in this direction and to consider, as a first approximation, a simplified linearized equation for vorticity.

Neglecting the nonlinear terms $(\boldsymbol{u} \cdot \nabla)\omega = u\,\partial\omega/\partial r + r^{-1} v\,\partial\omega/\partial\theta$ in the vorticity equation (2.1.22), we obtain an equation describing viscous diffusion of vorticity in a polar coordinate system

$$\frac{\partial \omega}{\partial t} = \nu \left(\frac{1}{r}\frac{\partial \omega}{\partial r} + \frac{\partial^2 \omega}{\partial r^2} + \frac{1}{r^2}\frac{\partial^2 \omega}{\partial \theta^2} \right). \tag{3.5.1}$$

This approximation is usually called the Stokes approximation, and the resulting flows are called Stokes flows. Obviously, solutions of the linear equation describe slow or 'weak' motions, i.e. flows where the Reynolds number is small. However, it turns out that, in spite of this essential simplification, it is possible to obtain some substantial results concerning the physical mechanism of vortex multipole generation in a viscous fluid, and to interpret experimental results on vortex dipole interactions (Chapter 4).

3.5.1 *Diffusion of a vortex multipole of arbitrary order*

Equation (3.5.1) states that the vorticity ω diffuses in the same way as some other physical properties (e.g. heat). Hence, to solve (3.5.1), we can use methods that have been developed for dealing with problems of heat transfer in solid media (e.g. Carslow and Jaeger, 1960).

In general, the unknown function ω depends on five independent governing parameters: A, ν, t, r and θ. Here A is a dimensional constant characterizing the amplitude of forcing and consequently the intensity of the flow. The dimensions of A are L^sT^p, where s and p are numbers determined by the type of forcing. Dimensional analysis shows that at least two quantities out of the five (e.g. ν and t) have independent dimensions. Therefore we can construct the non-dimensional combination

$$Re_* = A\nu^k t^m/4\pi, \tag{3.5.2}$$

where $k = -\frac{1}{2}s$ and $m = -(\frac{1}{2}s + p)$. The coefficient $1/4\pi$ is introduced here only for the sake of convenience. The non-dimensional amplitude of forcing Re_* may decrease ($m < 0$) or increase ($m > 0$) with time or be constant ($m = 0$). It can also be considered as the Reynolds number of the flow.

To reduce the number of arguments, let us introduce the non-dimensional similarity variable $\eta = \frac{1}{2}r(\nu t)^{-1/2}$. To determine the dependence of ω on the polar angle θ, let us use the multipole expansion (2.1.32), from which we expect the dependence of ω on θ to take the form

$$\omega \propto \cos[n(\theta + \theta_0)],$$

where θ_0 is some reference angle and $n = 0, 1, 2, \ldots$ is the order of the multipole.

Thus we are seeking self-similar solutions of (3.5.1) in the form (Barenblatt, Voropayev and Filippov, 1989)

$$\omega = Re_* \frac{1}{t} W(\eta)\cos[n(\theta + \theta_0)], \qquad \eta = \frac{r}{2(\nu t)^{1/2}}, \tag{3.5.3}$$

where Re_* is determined by (3.5.2) and W is a non-dimensional function. Substitution of (3.5.3) and (3.5.2) into (3.5.1) gives for $W(\eta)$

the ordinary differential equation

$$\frac{d^2W}{d\eta^2}+(2\eta+\eta^{-1})\frac{dW}{d\eta}-[n^2\eta^{-2}+4(m-1)]W=0. \tag{3.5.4}$$

By means of the substitution

$$W=x^{\pm n/2}\Phi(x), \qquad x=\eta^2, \tag{3.5.5}$$

(3.5.4) may be rewritten in the standard form

$$x\frac{d^2\Phi}{dx^2}+(x+1\pm n)\frac{d\Phi}{dx}-(m-1\mp\tfrac{1}{2}n)\Phi=0. \tag{3.5.6}$$

A general solution of (3.5.6) can be expressed in terms of confluent hypergeometric functions (e.g. Kamke, 1959). But these functions are too complicated for easy analysis. To obtain some solutions of (3.5.6) in terms of standard functions, let us consider some cases of practical interest.

To demonstrate how to use (3.5.6), we first repeat the solution obtained in section 3.2.1 for the diffusion of a point vortex of intensity Γ. The point vortex is a monopole; hence $n=0$. The dimensions of Γ are L^2T^{-1}, so that $s=2$ and $p=-1$. This gives $k=-1$ and $m=0$ in (3.5.2). For these values of n and m (3.5.6) becomes

$$x\frac{d^2\Phi}{dx^2}+(x+1)\frac{d\Phi}{dx}+\Phi=0.$$

Rewriting this in the form

$$\frac{d}{dx}\left(x\frac{d\Phi}{dx}+x\Phi\right)=0$$

and integrating it twice, we obtain the general solution of our problem

$$\Phi=C_1e^{-x}+C_2e^{-x}\int^x e^x x^{-1}\,dx, \tag{3.5.7}$$

where C_1 and C_2 are constants of integration. Assuming that there is no singularity at $r=0$ ($x=0$) for $t>0$, we set $C_2=0$, and we find for the vorticity

$$\omega=\frac{C_1\Gamma}{4\pi\nu t}e^{-r^2/4\nu t}. \tag{3.5.8}$$

Using the connection (2.1.29) between the total vorticity of the flow and the intensity Γ of the vortex monopole, we find $C_1 = 1$ in (3.5.8).

The stream function ψ of the flow satisfies the equation

$$\frac{\partial^2\psi}{\partial r^2} + \frac{1}{r}\frac{\partial\psi}{\partial r} + \frac{1}{r^2}\frac{\partial^2\psi}{\partial\theta^2} = -\omega. \tag{3.5.9}$$

Substituting (3.5.8) into (3.5.9) and solving the resulting equation, we find for the stream function

$$\psi = \frac{C_1\Gamma}{2\pi}\left(\frac{1}{2}\int^{r^2/4\nu t} e^{-x}x^{-1}\,dx - C_3\log r + c\right), \tag{3.5.10}$$

where c is some inessential additive constant. The second constant C_3 is determined from the appropriate initial conditions at $t = 0$. For a point vortex

$$v(0, r) = -\frac{\partial\psi(0, r)}{\partial r} = \frac{\Gamma}{2\pi r},$$

which gives $C_3 = 1$, while for the flow induced by a rotating needle the initial and boundary conditions are

$$v(0, r) = 0 \quad (r > 0),$$
$$2\pi r v(r, t) = \Gamma \quad (r = 0),$$

which give $C_3 = 0$ and $C_1 = -1$.

Thus we have repeated the solutions (3.2.4) and (3.2.7) for the monopoles, and we now seek self-similar Stokes solutions in the form of (3.5.3) for higher-order multipoles.

3.5.2 The Stokes dipole

Let us first formulate the appropriate initial condition for this problem. In principle, the velocity distribution, given by (3.1.12) for a point vortex dipole in an ideal fluid, can be used for this purpose. Then, 'turning on' the viscosity at $t = 0$, one can study the viscous evolution of the flow with time. However, in this case it remains unclear how such initial motion can be created in practice. On the other hand, we know that the dipole moment or the impulse of the vorticity distribution given by (2.1.30) is equal to I, where ρI is the total impulse of the flow. Thus, imparting locally (in the ideal case at a point) a momentum ρI to the fluid, we can reproduce in a viscous fluid the same initial flow as that induced by a point vortex dipole of intensity I. In experiments a thin nozzle from which a fluid is injected can be used as the point source of momentum. The source may act in different regimes: impulsively, imparting total momentum ρI during a short period of time, or with constant intensity ρJ. In the latter case the total momentum of the flow and hence the intensity of the dipole increase linearly with time: $I = Jt$.

Thus let us consider the flow induced by a point momentum source that acts impulsively at time $t=0$ and exerts on the fluid a force

$$\rho J = \rho I \delta(t)$$

in the direction $\theta = 0$. Here $\delta(t)$ is the Dirac delta function. The source is at the point $r=0$. The dimensions of I are L^3T^{-1}. In this case $s=3$ and $p=1$, and in (3.5.2) we have $k=-\frac{3}{2}$ and $m=-\frac{1}{2}$. For a dipole $n=1$. From the direction of action of the source we can conclude that $\theta_0 = \frac{1}{2}\pi$ in (3.5.3). According to (3.5.3), we seek a solution in the form

$$\omega = Re_* t^{-1} W(\eta) \sin\theta, \qquad Re_* = (4\pi)^{-1} I \nu^{-3/2} t^{-1/2}.$$

Choosing the minus sign in the exponent in (3.5.5), we can represent (3.5.6) in the simple form

$$\frac{\mathrm{d}}{\mathrm{d}x}\left(x\frac{\mathrm{d}\Phi}{\mathrm{d}x} + x\Phi - \Phi\right) = 0, \qquad \Phi = x^{1/2}W, \qquad x = \eta^2.$$

Integrating this equation, we obtain the general solution for W in the form

$$W(\eta) = C_1 \eta \mathrm{e}^{-\eta^2} + C_2 \eta \mathrm{e}^{-\eta^2} \int^{\eta^2} \mathrm{e}^x x^{-2}\,\mathrm{d}x.$$

From the formulation of the problem we assume that there is no singularity at $r=0$ ($\eta = 0$) for $t>0$; so that $C_2 = 0$, which gives

$$\omega = \frac{C_1 I r}{8\pi(\nu t)^2}\mathrm{e}^{-r^2/4\nu t}\sin\theta. \tag{3.5.11}$$

To find the constant C_1, we use the relation (3.4.5) between the impulse I applied by the force distribution since the onset of the motion and the impulse of the vorticity distribution. This relation may be written in the form

$$I = \int_0^{2\pi} \int_0^{\infty} \sin\theta'\, \omega(r', \theta') r'^2 \,\mathrm{d}r'\,\mathrm{d}\theta'.$$

For the vorticity given by (3.5.11) this gives $C_1 = 1$. Thus we finally obtain

$$\omega = \frac{Ir}{8\pi(\nu t)^2} \mathrm{e}^{-r^2/4\nu t} \sin\theta. \tag{3.5.12}$$

After solving (3.5.9) we find the stream function of the flow:

$$\psi = \frac{I}{2\pi r}(1 - \mathrm{e}^{-r^2/4\nu t}) \sin\theta. \tag{3.5.13}$$

Now consider the case in which a point momentum source acts continuously, starting at time $t = 0$ and thereafter exerting on the fluid a force $J\rho$ in the direction $\theta = 0$. The dimensions of J are L^3T^{-2}, and $J = \text{const}$. Using the same arguments as above and taking into account that for this case $k = -\frac{3}{2}$, $m = \frac{1}{2}$, $\theta_0 = \frac{1}{2}\pi$ and $n = 1$, we obtain

$$\omega = \frac{J}{2\pi\nu r} \mathrm{e}^{-r^2/4\nu t} \sin\theta, \tag{3.5.14}$$

$$\psi = \frac{Jr}{8\pi\nu}\left[\frac{4\nu t}{r^2}(1 - \mathrm{e}^{-r^2/4\nu t}) + \int_{r^2/4\nu t}^{\infty} \mathrm{e}^{-x} x^{-1}\,\mathrm{d}x\right] \sin\theta. \tag{3.5.15}$$

The distributions of vorticity across the axis of the flow given by (3.5.14) for different times are shown in Figure 3.7.

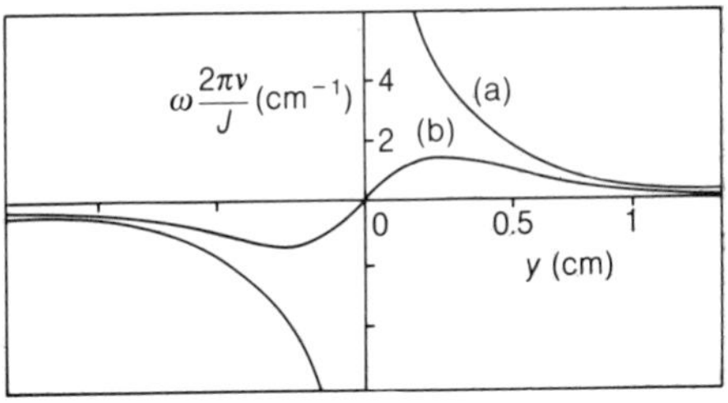

Figure 3.7 *Distribution according to (3.5.14) of the normalized vorticity in a vortex dipole across the x axis ($\theta = 0$) in Cartesian coordinates (x, y) for $t = 25$ s and $\nu = 10^{-2}$ cm^2 s^{-1}: (a) $x = 0$; (b) 0.3 cm.*

Although these self-similar Stokes solutions are obtained in the linear approximation, they are very helpful for better understanding of real flow evolution. Integrating the equations of motion for marked particles

$$\left.\begin{aligned}\frac{\mathrm{d}r}{\mathrm{d}t}&=\frac{1}{r}\frac{\partial\psi}{\partial\theta},\\ \frac{\mathrm{d}\theta}{\mathrm{d}t}&=-\frac{1}{r}\frac{\partial\psi}{\partial r},\end{aligned}\right\}\tag{3.5.16}$$

with the stream function (3.5.13) or (3.5.15), one can calculate the trajectory of a marked particle starting at $t=t_0$ from the point (r_0, θ_0). Typical results of such a calculation for the case when the source of constant intensity J acts continuously, thus inducing a dipolar flow whose impulse ρI increases linearly with time ($I=Jt$), are shown in Figure 3.8. The equations of motion for marked particles are obtained by substituting (3.5.15) into (3.5.16), giving

$$\frac{\mathrm{d}r}{\mathrm{d}t}=\frac{J}{8\pi\nu}\left[\frac{4\nu t}{r^2}(1-\mathrm{e}^{-r^2/4\nu t})+\int_{r^2/4\nu t}^{\infty}\mathrm{e}^{-x}x^{-1}\,\mathrm{d}x\right],\tag{3.5.17}$$

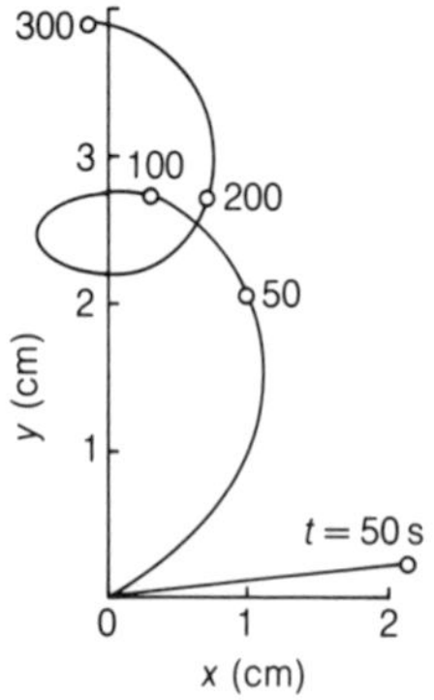

Figure 3.8 *Typical trajectories of two 'marked' particles in a vortex dipole flow in Cartesian coordinates (x, y). The points show the positions of the particles at different times t for $\nu = 10^{-2}\,\mathrm{cm}^2\,\mathrm{s}^{-1}$ and $J = 0.02\,\mathrm{cm}^3\,\mathrm{s}^{-2}$. The particles start at $t = 0$ from different points in the vicinity of the origin.*

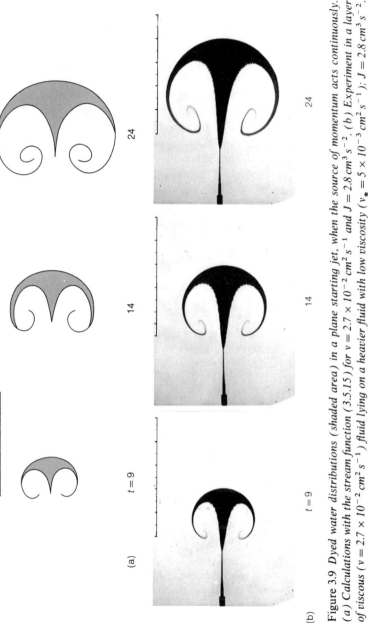

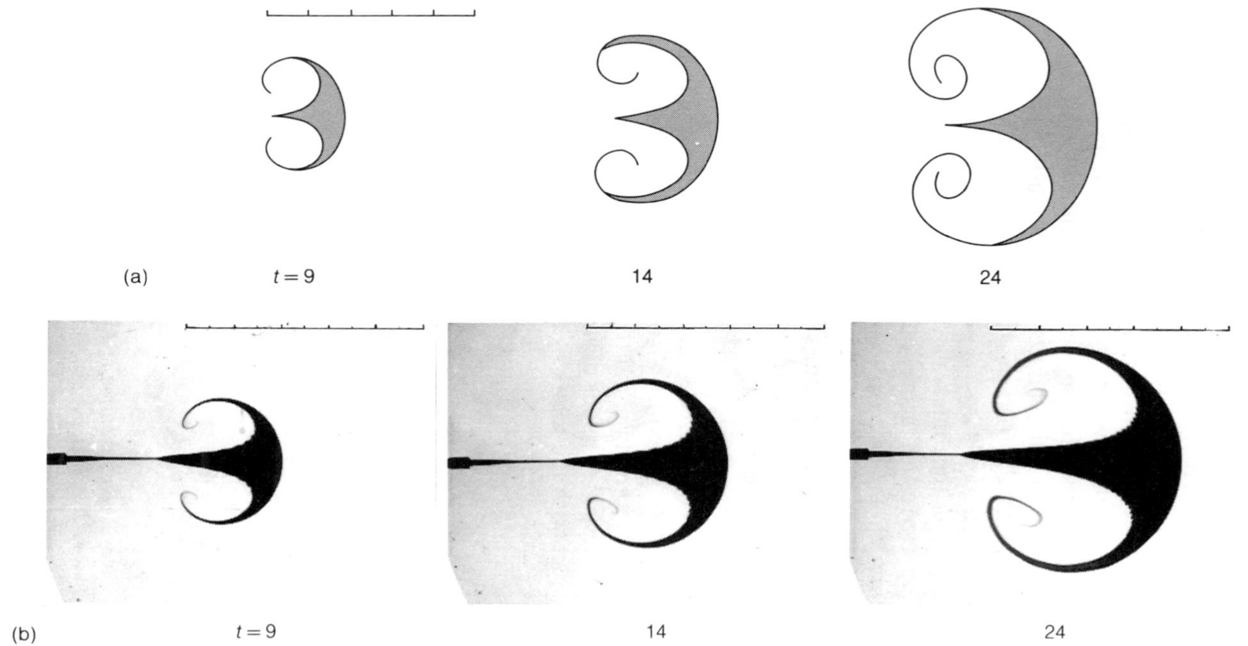

Figure 3.9 *Dyed water distributions (shaded area) in a plane starting jet, when the source of momentum acts continuously. (a) Calculations with the stream function (3.5.15) for* $\nu = 2.7 \times 10^{-2}\ cm^2\ s^{-1}$ *and* $J = 2.8\ cm^3\ s^{-2}$. *(b) Experiment in a layer of viscous* ($\nu = 2.7 \times 10^{-2}\ cm^2\ s^{-1}$) *fluid lying on a heavier fluid with low viscosity* ($\nu_* = 5 \times 10^{-3}\ cm^2\ s^{-1}$); $J = 2.8\ cm^3\ s^{-2}$. *Scales are in centimetres and time t in seconds.*

$$\frac{d\theta}{dt} = \frac{J}{8\pi\nu r}\left[\frac{4\nu t}{r^2}(1 - e^{-r^2/4\nu t}) - \int_{r^2/4\nu t}^{\infty} e^{-x}x^{-1}\,dx\right]. \quad (3.5.18)$$

In Figure 3.8 one can see that particles that are initially close together may describe different trajectories. To obtain more information about the whole flow, one must analyse the motion of a large number of particles. This can be done in the following way.

In experiments a dyed fluid is injected continuously from a thin slitted nozzle of width $d = 0.02\,\text{cm}$. In computations the marked particles are injected discretly from several points on the arc $r_0 = \frac{1}{2}d$, $-\frac{1}{2}\pi \leqslant \theta \leqslant \frac{1}{2}\pi$ at small time intervals. The endpoints of all injected particles at time t give the theoretical distributions of dyed fluid shown in Figure 3.9(a). For comparison, photographs of the real flow in a thin layer of viscous fluid are shown in Figure 3.9(b). In spite of the essential simplifications employed, the linear solution correctly reproduces the general features of the flow – its spiral mushroom-like structure – and is in quite good agreement with observations.

Thus the action of a localized force on a viscous fluid induces a compact vortex dipole structure. Figure 3.10 shows schematically the process of dipole formation and the corresponding distribution of dyed fluid, as well as the regions of vortical and nearly irrotational flow in accordance with the solutions obtained.

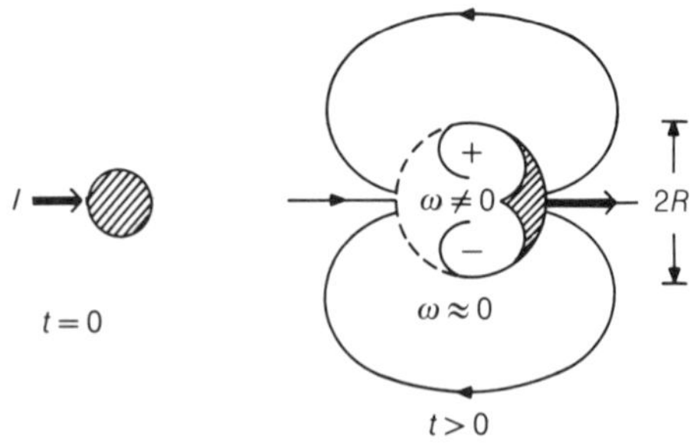

Figure 3.10 *Sketch of dipole formation under the action of a localized force: appropriate dyed water distributions (shaded area), vortical flow ($\omega \neq 0$) and irrotational flow ($\omega = 0$).*

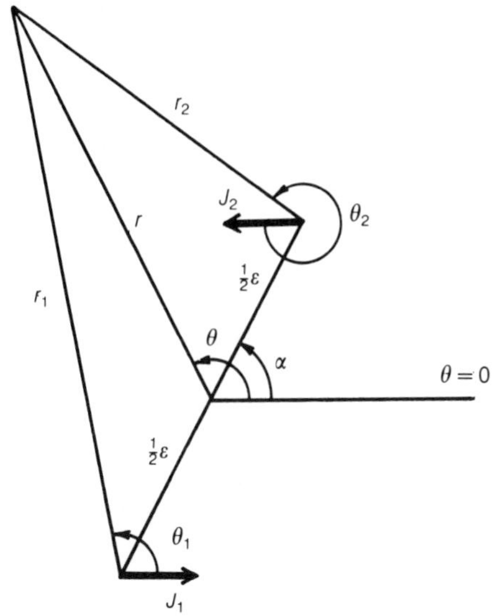

Figure 3.11 *Sketch of two equal ($J_1 = J_2 = J$) forces acting on a fluid in opposite directions: $\theta = \pi$ and $\theta = 0$. α is an arbitrary angle.*

3.5.3 *The Stokes quadrupole*

Consider two point sources of momentum exerting equal forces $J\rho$ on a fluid in opposite directions (Figure 3.11). The sources are at the points $(\frac{1}{2}\varepsilon, \alpha)$ and $(\frac{1}{2}\varepsilon, \pi + \alpha)$, and act on the fluid in the directions $\theta = \pi$ and $\theta = 0$ respectively; α is an arbitrary angle. On decreasing ε and increasing J such that the product $\varepsilon J = Q$ remains constant, we obtain a point source of intensity Q at the point $r = 0$. The source consisting of two forces (Figure 3.11) is a force dipole for $\alpha = 0, \pi$, and a pair of forces for $\alpha = \pm\frac{1}{2}\pi$. For arbitrary α we have a combination of the two types of source. The force dipole applies no net force on the fluid. Hence the dipolar distribution of vorticity (2.1.30) is zero and the first non-zero term in (2.1.28) and (2.1.32) is the one with quadrupolar vorticity distribution (2.1.31); therefore $\omega \propto \cos 2(\theta + \theta_0)$.

Consider again the two cases of an impulsive source and a source of constant intensity. In the first $Q = M\delta(t)$, where the dimensions of M are L^4T^{-1}. Hence in (3.5.2), (3.5.5) and (3.5.6) $k = -2$, $m = -1$ and

$n = 2$, and on solving (3.5.5) and (3.5.6) we obtain a solution satisfying the requirement that $W(\eta) \to 0$ as $\eta \to \infty$:

$$W(\eta) = C\eta^2 \mathrm{e}^{-\eta^2}. \qquad (3.5.19)$$

For a source of constant intensity Q that starts at $t = 0$ the dimensions of Q are $L^4 T^{-2}$. Hence $k = -2$, $m = 0$ and $n = 2$, and we obtain a solution in the form

$$W(\eta) = C(1 + \eta^{-2})\mathrm{e}^{-\eta^2}. \qquad (3.5.20)$$

To determine the constant of integration C in (3.5.19) and (3.5.20), we need a relation, similar to (2.1.30), between the source intensity and the appropriate quadrupolar vorticity distribution. We must also find some additional arguments to determine the constant θ_0 in (3.5.3).

For large r the stream function (3.5.13) or (3.5.15) for the flow induced by a point force is

$$\psi_0 = \frac{I}{2\pi r} \sin \theta, \qquad (3.5.21)$$

where $I = \text{const}$ for an impulsive force and $I = Jt$ for a constant force. Representing the stream function for two equal forces ($J_1 = J_2 = J$), shown in Figure 3.11 as the superposition of stream functions (3.5.21), for large r, we have

$$\psi = \frac{I}{2\pi}\left(\frac{\sin \theta_1}{r_1} + \frac{\sin \theta_2}{r_2}\right). \qquad (3.5.22)$$

From the geometry (Figure 3.11) we have

$$\begin{aligned} r_1^2 &= r^2 + (\tfrac{1}{2}\varepsilon)^2 + \varepsilon r \cos(\theta - \alpha), \\ r_2^2 &= r^2 + (\tfrac{1}{2}\varepsilon)^2 - \varepsilon r \cos(\theta - \alpha), \\ r_1 \sin \theta_1 &= r \sin \theta + \tfrac{1}{2}\varepsilon \sin \alpha, \\ r_2 \sin \theta_2 &= -r \sin \theta + \tfrac{1}{2}\varepsilon \sin \alpha, \end{aligned}$$

and for $r \gg \frac{1}{2}\varepsilon$, with an accuracy of order $\varepsilon/2r$, we obtain for the force dipole

$$\psi = -\frac{M}{2\pi r^2} \sin(2\theta - \alpha), \qquad (3.5.23)$$

where $M = \varepsilon I$.

Comparing (3.5.23) with the third terms in the expansions (2.1.28) and (2.1.32), we obtain the condition determining the constants C and θ_0

$$-M\sin(2\theta-\alpha)=\tfrac{1}{2}\int_0^{2\pi}\int_0^{\infty}\cos 2(\theta-\theta')\,\omega(r',\theta',t)r'^3\,\mathrm{d}r'\,\mathrm{d}\theta'. \quad (3.5.24)$$

Using this condition, after some algebra, we find $C=1/8\pi$ and $\theta_0=-\frac{1}{2}\alpha-\frac{1}{4}\pi$. Inserting these constants into (3.5.19), (3.5.20) and (3.5.3) and solving (3.5.9), we obtain two solutions. For the impulsive source, $Q=M\delta(t)$, we have

$$\omega=-\frac{Q}{8\pi\nu^2 t}\left(1+\frac{4\nu t}{r^2}\right)\mathrm{e}^{-r^2/4\nu t}\sin(2\theta-\alpha), \quad (3.5.27)$$

$$\psi=-\frac{Qt}{2\pi r^2}(1-\mathrm{e}^{-r^2/4\nu t})\sin(2\theta-\alpha). \quad (3.5.28)$$

For the source of constant intensity, $M=Qt$, we have

$$\omega=-\frac{Q}{8\pi\nu^2 t}\left(1+\frac{4\nu t}{r^2}\right)\mathrm{e}^{-r^2/4\nu t}\sin(2\theta-\alpha), \quad (3.5.27)$$

$$\psi=-\frac{Qt}{2\pi r^2}(1-\mathrm{e}^{-r^2/4\nu t})\sin(2\theta-\alpha). \quad (3.5.28)$$

By integrating the equations of motion (3.5.16) with the stream function (3.5.26) or (3.5.28), one can calculate the distributions of marked particles that were initially in a small circle around the source. These distributions are shown for the impulsive force dipole (3.5.26) in Figure 3.12.

Thus two equal forces acting in opposite directions induce a vortex quadrupole consisting of two dipoles moving in the directions $\theta=\frac{1}{2}(\alpha\pm\pi)$. Vortex quadrupole formation and the appropriate distributions of dyed fluid, vortical and irrotational flows are shown schematically in Figure 3.13.

Note that, because the solutions obtained are linear ones, a simple superposition of two or more solutions gives a new linear solution. Consider for example the case where two equal forces act in opposite directions continuously and the distance ε between the points where the forces are applied is finite and does not tend to zero. The

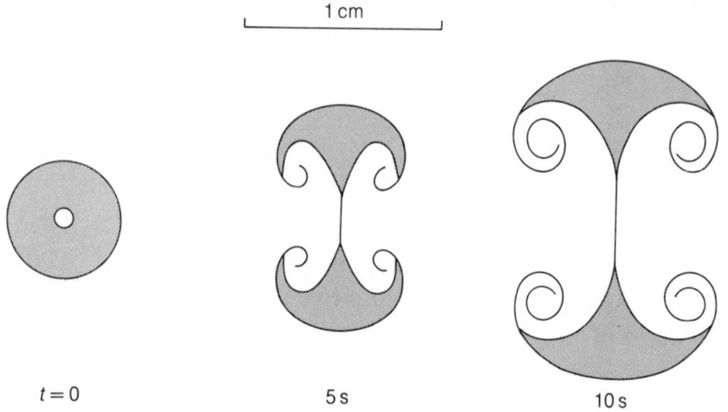

Figure 3.12 *Dyed water distribution (shaded area) in a planar impulsive vortex quadrupole: calculations using the stream function (3.5.26) at* $M = 2\,\mathrm{cm}^4\,\mathrm{s}^{-1}$ *and* $\nu = 10^{-2}\,\mathrm{cm}^2\,\mathrm{s}^{-1}$. *The contours of the shaded area were obtained by calculating the trajectories of 10 marked particles in the first quadrant. The final positions of the particles for different times were drawn by hand. To avoid the singularity, particles with initial positions near the point* $r = 0$ *were not considered.*

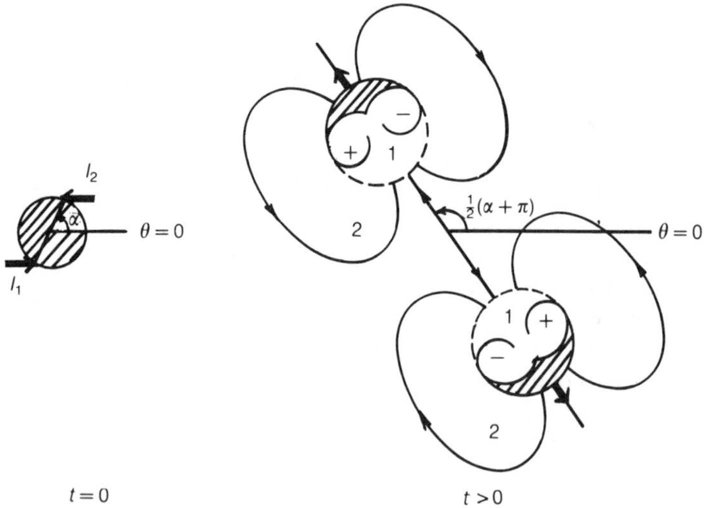

Figure 3.13 *Sketch of quadrupole formation under the action of two equal forces of opposite directions: appropriate dyed water distributions (shaded area), vortical flow (1) and potential flow (2).*

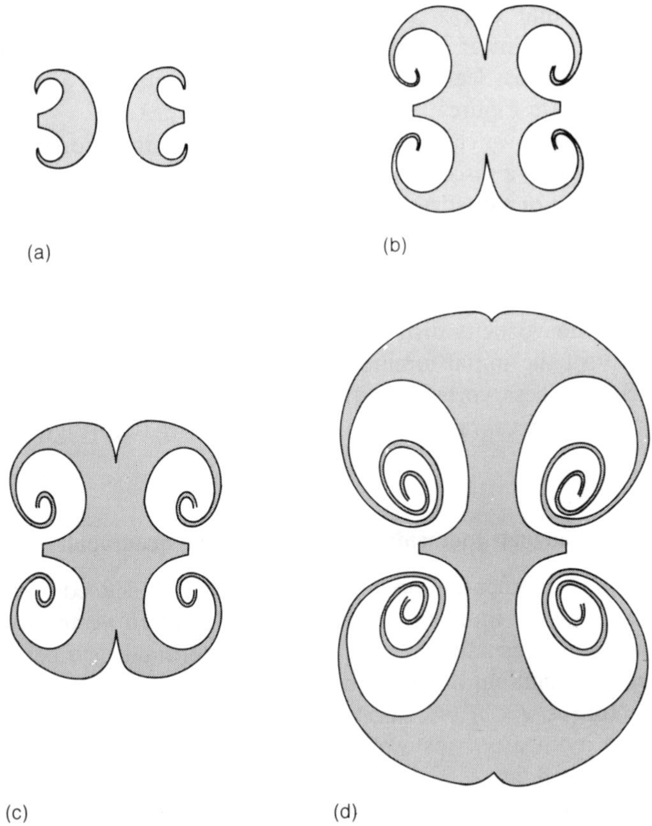

Figure 3.14 *A mathematical model of the symmetric collision of two dipoles. Dyed water distributions in the flow were calculated using the superposition of two stream functions (3.5.15). The distance between sources* $\varepsilon = 1.0$ *cm, the force amplitude* $J = 0.04\,cm^3\,s^{-2}$, *and* $\nu = 10^{-2}\,cm^2\,s^{-1}$. *The final positions of 72 marked particles are contoured at times* $t = 3$ *s (a), 7 s (b), 9 s (c) and 25 s (d).*

appropriate superposition of two stream functions (3.5.15) gives the resulting flow. The subsequent evolution of the distribution of the dyed fluid is shown in Figure 3.14. In the initial stage of flow evolution two dipoles induced by each force develop almost independently (Figure 3.14a). As time progresses, the dipoles begin to interact (Figure 3.14b), and finally form a vortex quadrupole (Figure 3.14c,d), similar to that shown in Figure 3.12. Although it might seem surprising, the linear solution correctly reproduces (at least qualitatively) the gross features of the real flow evolution in a stratified fluid (compare Figures 1.5, 1.7 and Figures 3.12, 3.14).

In summary, we can say that the action of a single force or two equal forces on a viscous fluid leads to the formation of a compact vortex dipole or a quadrupole respectively. The governing parameter for the dipole is the integral intensity I of the dipolar vorticity distribution, which is equal to the impulse of the dipole. Correspondingly, for the quadrupole it is the total intensity M of the quadrupolar vorticity distribution. I and M are determined by the intensity of the initial forcing on the fluid. Note, finally, that the vorticity in these vortex structures is localized in a compact fluid volume of a typical radius R, which increases with time (Figures 3.10 and 3.13).

3.6 Flow angular momentum and the rotating quadrupole

The angular momentum per unit volume of fluid is defined as $\boldsymbol{x} \times \rho \boldsymbol{u}$ where $\rho \boldsymbol{u}$ is the momentum and $\boldsymbol{x}$ is the position vector from a chosen point (usually the origin of the coordinate system). The angular momentum appears and changes as a result of the action of the torque $\boldsymbol{x} \times \rho \boldsymbol{f}$ of external forces on a fluid. Conservation of angular momentum (as well as conservation of momentum) is assumed for fluids.

In the problem of the flow induced by two opposite forces (Figure 3.11), in addition to a force dipole of intensity $\rho Q = \rho \varepsilon J$, a torque $\rho \mathcal{M} = \rho Q \sin \alpha$ is applied to the fluid when $\alpha \neq 0$. One expects that the induced flow will possess non-zero angular momentum. However, the solutions (3.5.26) and (3.5.28) do not satisfy this requirement, and the total angular momentum ρN of the flow is identically zero:

$$N = \int_S \boldsymbol{x} \times \boldsymbol{u} \, \mathrm{d}S = \int_0^{2\pi} \int_0^{\infty} r^2 v \, \mathrm{d}r \, \mathrm{d}\theta = 0,$$

since

$$v = -\frac{\partial \psi}{\partial r} \propto \sin(2\theta - \alpha).$$

Thus these solutions fail to describe some details of the flow. To resolve this contradiction, let us consider in greater detail the dynamical conditions of conservation of angular momentum and its relation to the appropriate vorticity distribution.

3.6.1 Conservation of angular momentum

Angular momentum conservation was used implicitly in the derivation of the momentum conservation equation (2.1.13), when the symmetry of the tensor Π_{ik} was assumed. The cross-product of $\boldsymbol{x}$ and the momentum equation with the external forces on the right-hand side,

$$\frac{\partial}{\partial t}\rho u_i + \frac{\partial}{\partial x_k}\Pi_{ik} = \rho f_i, \tag{3.6.1}$$

gives

$$\frac{\partial}{\partial t}(\boldsymbol{x}\times\rho\boldsymbol{u})_l + \varepsilon_{lji}\left(\frac{\partial}{\partial x_k}x_j\Pi_{ik} - \Pi_{ij}\right) = (\boldsymbol{x}\times\rho\boldsymbol{f})_l, \tag{3.6.2}$$

where the tensor ε_{lji} has been used to write the second term and $\partial x_i/\partial x_k = \delta_{ik}$ has been taken into account. Integrating (3.6.2) over a fixed arbitrary volume V bounded by a surface S,

$$\frac{d}{dt}\int_V (\boldsymbol{x}\times\rho\boldsymbol{u})_l\,dV + \varepsilon_{lji}\int_S x_j\Pi_{ik}\,dS_k - \varepsilon_{lji}\int_V \Pi_{ij}\,dV = \int_V (\boldsymbol{x}\times\rho\boldsymbol{f})_l\,dV$$

we obtain, after summation over i and j, for the $x_3 = z$ component

$$\frac{d}{dt}\int_V (\boldsymbol{x}\times\rho\boldsymbol{u})_z\,dV + \int_S [\boldsymbol{x}\times(\Pi_{ik}\,dS_k)]_z - \int_V (\Pi_{xy} - \Pi_{yx})\,dV$$
$$= \int_V (\boldsymbol{x}\times\rho\boldsymbol{f})_z\,dV.$$

For a symmetrical tensor $\Pi_{ik} = \Pi_{ki}$ the second volume integral is zero. A similar procedure can be carried out for the other components. Finally, one obtains the angular momentum balance in integral form:

$$\frac{d}{dt}\int_V (\boldsymbol{x}\times\rho\boldsymbol{u})\,dV = -\int_S \boldsymbol{x}\times(\Pi_{ik}\,dS_k) + \int_V (\boldsymbol{x}\times\rho\boldsymbol{f})\,dV, \tag{3.6.3}$$

where $\Pi_{ik}\,\mathrm{d}S_k$ is the ith momentum component through the surface element normal to the x_k axis. Thus the rate of change of angular momentum of fluid contained in a volume fixed in space is equal to the total torque of external forces applied to the fluid in this volume minus the outward flux of angular momentum through the boundaries. For a planar flow the angular momentum vector is directed along the normal to the plane of the flow, and in polar coordinates we have

$$\frac{\mathrm{d}}{\mathrm{d}t}\int_S \rho r v\,\mathrm{d}S = -R^2\int_L \Pi_{r\theta}\,\mathrm{d}\theta + \int_S \rho r f_\theta\,\mathrm{d}S, \tag{3.6.4}$$

where f_θ is the azimuthal component of the external forces, L is a circle of radius R and S is the region inside this circle.

It turns out that the total angular momentum of the flow is connected with the vorticity distribution. This connection can be found in a similar way as that between the flow momentum (3.4.2) and the impulse (3.4.5) of the vorticity distribution (the first moment of the vorticity distribution).

For a planar flow (Figure 2.1) the kinematic angular momentum per unit depth is expressed as

$$N(t) = \int_S (\boldsymbol{x}\times\boldsymbol{u})\,\mathrm{d}S. \tag{3.6.5}$$

Using the identity

$$\boldsymbol{x}\times\boldsymbol{u} = \boldsymbol{x}\times(\nabla\times\boldsymbol{B}) = \nabla\times(\boldsymbol{x}\times\boldsymbol{B}) + 2\boldsymbol{B},$$

where $\boldsymbol{B} = (0, 0, \psi)$ is the vector potential and $\nabla\cdot\boldsymbol{B} = 0$, we rewrite (3.6.5) in the form

$$N(t) = \int_L \boldsymbol{n}\times(\boldsymbol{x}\times\boldsymbol{B})\,\mathrm{d}L + 2\int_S \boldsymbol{B}\,\mathrm{d}S. \tag{3.6.6}$$

Substituting $\boldsymbol{B}$ from (3.4.1) into (3.6.6), exchanging the order of integration, and using

$$\int_L \log|\boldsymbol{x}-\boldsymbol{x}'|\,\mathrm{d}\theta = 4\pi\times\begin{cases}\log|\boldsymbol{x}| & (|\boldsymbol{x}|\geqslant|\boldsymbol{x}'|),\\ \log|\boldsymbol{x}'| & (|\boldsymbol{x}|\leqslant|\boldsymbol{x}'|),\end{cases}$$

$$\boldsymbol{n}\times(\boldsymbol{x}\times\boldsymbol{B}) = \boldsymbol{x}(\boldsymbol{n}\cdot\boldsymbol{B}) - (\boldsymbol{n}\cdot\boldsymbol{x})\boldsymbol{B},$$

leads to

$$\int_L \boldsymbol{n} \times (\boldsymbol{x} \times \boldsymbol{B})\,\mathrm{d}L = R^2 \log R \int_{S'} \boldsymbol{\omega}(\boldsymbol{x}', t)\,\mathrm{d}S', \tag{3.6.7}$$

$$2\int_S \boldsymbol{B}\,\mathrm{d}S = -R^2(\log R - \tfrac{1}{2}) \int_{S'} \boldsymbol{\omega}(\boldsymbol{x}', t)\,\mathrm{d}S' - \tfrac{1}{2}\int_{S'} |\boldsymbol{x}'|^2 \boldsymbol{\omega}(\boldsymbol{x}', t)\,\mathrm{d}S', \tag{3.6.8}$$

where $R = |\boldsymbol{x}|$ and $\mathrm{d}L = R\,\mathrm{d}\theta$. Addition of (3.6.7) and (3.6.8) gives

$$\boldsymbol{N}(t) = \int_{S'} (\boldsymbol{x} \times \boldsymbol{u})\,\mathrm{d}S = \tfrac{1}{2}\int_S \boldsymbol{x} \times (\boldsymbol{x} \times \boldsymbol{\omega})\,\mathrm{d}S + \tfrac{1}{2}R^2 \int_S \boldsymbol{\omega}(\boldsymbol{x}, t)\,\mathrm{d}S. \tag{3.6.9}$$

When the total vorticity is zero,

$$\boldsymbol{\Gamma} = \int_S \boldsymbol{\omega}(\boldsymbol{x}, t)\,\mathrm{d}S = 0, \tag{3.6.10}$$

one obtains

$$\boldsymbol{N}(t) = \tfrac{1}{2}\int_S \boldsymbol{x} \times (\boldsymbol{x} \times \omega)\,\mathrm{d}S. \tag{3.6.11}$$

In the opposite case the angular momentum becomes infinitely large as $R \to \infty$.

In practice, it is the variation of the total angular momentum that is essential for the flow evolution rather than its absolute value. That is why one can subtract the stream function

$$\psi_0 = \frac{\Gamma}{2\pi} \log |\boldsymbol{x}|$$

from the stream function of the considered flow with $\boldsymbol{\Gamma} \neq 0$ so that the angular momentum of the resulting flow will be finite. A simple example is the diffusion of a point vortex with velocity distribution $v(r, t)$ given by (3.2.4). Although the total angular momentum for the original flow is infinite, the difference

$$\int_0^{2\pi}\int_0^{\infty} \left[v(r, t) - \frac{\Gamma}{2\pi r} \right] r^2\,\mathrm{d}r\,\mathrm{d}\theta$$

is finite and equal to

$$-\frac{1}{2}\int_0^{2\pi}\int_0^{\infty} r^3\omega(r,t)\,\mathrm{d}r\,\mathrm{d}\theta = -2\nu t\Gamma.$$

For the introduced additional flow, the first term in the expansion (2.1.28) is zero, so that $\boldsymbol{u} \sim 1/R^2$ for large distances $|\boldsymbol{x}| = R$. Thus for the component of the momentum flux tensor in (3.6.4),

$$\Pi_{r\theta} = \rho uv - \nu\rho\left[r\frac{\partial}{\partial r}\left(\frac{v}{r}\right) + \frac{1}{r}\frac{\partial u}{\partial \theta}\right],$$

we find the following estimate as $R \to \infty$:

$$\Pi_{r\theta} \sim \frac{1}{R^3} + O\left(\frac{1}{R^4}\right).$$

As $R \to \infty$, the contour integral in (3.6.4) tends to zero, and we have

$$N(t) = \int_0^t \int_S \boldsymbol{x} \times \boldsymbol{f}(\boldsymbol{x}, t)\,\mathrm{d}S\,\mathrm{d}t. \tag{3.6.12}$$

Thus the torque applied by external forces goes entirely into the creation of the angular momentum of the flow. Recall that one-half of the impulse supplied by the external forces is removed by the pressure field at infinity, which opposes the motion, and only the remaining half ends up in the momentum of the fluid. This result also holds for three-dimensional flows, with the only difference lying in the values of the constant factors, which are equal to 1/2 in (3.4.5) and 1/3 in (3.4.4) and (3.6.11).

3.6.2 *The rotating quadrupole*

To illustrate the connection (3.6.11) between the angular momentum of the flow and the second moment of the vorticity distribution, let us consider a simple example. Let the initial distribution of the azimuthal velocity be given by

$$v = \begin{cases} r\Omega & (r < a), \quad \Omega = \text{const} \\ 0 & (r > a). \end{cases} \tag{3.6.13}$$

We can rewrite (3.6.13) in the more convenient form

$$v = r\Omega[1 - H(r - a)], \tag{3.6.14}$$

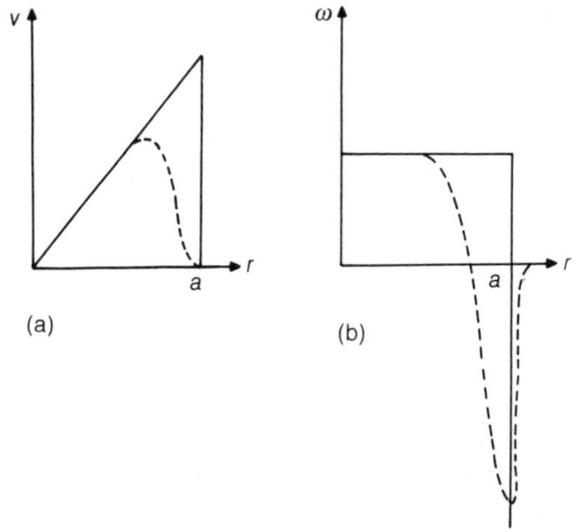

Figure 3.15 *Radial distributions of the azimuthal velocity v and vorticity ω for a localized monopole. Solid lines represent the expressions (3.6.14) and (3.6.15); dashed lines show the appropriate real distributions.*

using the Heaviside step function

$$H(x) = \begin{cases} 0 & (x < 0), \\ \frac{1}{2} & (x = 0), \\ 1 & (x > 0). \end{cases}$$

The corresponding vorticity distribution is shown schematically in Figure 3.15, and is given by

$$\omega = \frac{1}{r}\frac{\partial}{\partial r} rv = \Omega\{2[1 - H(r-a)] - r\delta(r-a)\}, \qquad (3.6.15)$$

where $\delta(x)$ is the Dirac delta function, with

$$\frac{\mathrm{d}}{\mathrm{d}x} H(x) = \delta(x).$$

The angular momentum ρN of the flow, with

$$N = \int_0^{2\pi} \int_0^{\infty} vr^2 \, \mathrm{d}r \, \mathrm{d}\theta = \tfrac{1}{2}\pi\Omega a^4,$$

is equal to the second moment of the vorticity distribution,

$$-\frac{1}{2}\int_0^{2\pi}\int_0^{\infty}\omega r^3\,\mathrm{d}r\,\mathrm{d}\theta = -\pi\Omega(\tfrac{1}{2}a^4 - a^4) = \tfrac{1}{2}\pi\Omega a^4, \tag{3.6.16}$$

while the total vorticity is zero,

$$\Gamma = \int_0^{2\pi}\int_0^{\infty}\omega r\,\mathrm{d}r = 2\pi\Omega(a^2 - a^2) = 0.$$

In practice, similar initial distributions can be created in the following way. A bottomless cylinder with thin walls is placed in the fluid, and the fluid inside the cylinder is rotated in some way (while the cylinder remains still) until solid-body rotation is established. Then the cylinder is carefully withdrawn. The resulting distributions are shown schematically in Figure 3.15. The vortex obtained is also called an isolated vortex monopole. For this flow all terms in the expansion (2.1.28) are identically zero, so the fluid is at rest for $r > a$. Only at the moment of creation of the vortex (when the cylinder is withdrawn) does a radial pressure gradient appear, as a result of the action of centrifugal force. The subsequent viscous diffusion of such a vortex was studied by Nekrasov (1931) (see also Kochin, Kibel and Rose, 1963).

It is clear that the additional terms in the solutions (3.5.26) and (3.5.28) for quadrupoles, when $\alpha \neq 0$ and angular momentum is transported to the fluid, must describe an isolated monopole; hence $n = 0$ in (3.5.3). For an impulsive quadrupole the intensity of the angular momentum source is $Q\sin\alpha = M\delta(t)\sin\alpha$. The dimensions of M are L^4T^{-1}, which gives $k = -2$ and $m = -1$ in (3.5.2). From (3.5.6) we find

$$x\frac{\mathrm{d}^2\Phi}{\mathrm{d}x^2} + (1 + x)\frac{\mathrm{d}\Phi}{\mathrm{d}x} + 2\Phi = 0. \tag{3.6.17}$$

The general solution of (3.6.17) can be written as

$$\Phi = C(1 - x)\mathrm{e}^{-x} + C_1(1 - x)\mathrm{e}^{-x}\int^x \frac{\mathrm{e}^y\,\mathrm{d}y}{y(1 - y)^2}.$$

Since there is no singularity at the origin $r = 0$ ($x = 0$) when $t > 0$, we find $C_1 = 0$, so that ω in (3.5.3) becomes

$$\omega = \frac{CM \sin\alpha}{4\pi(\nu t)^2}(1-x)\,\mathrm{e}^{-x}, \qquad x = \frac{r^2}{4\nu t} \tag{3.6.18}$$

Substituting (3.6.18) into (3.6.11) for

$$N(t) = \sin\alpha \int_0^t M\delta(t)\,\mathrm{d}t = M\sin\alpha,$$

we have $C = \frac{1}{2}$. The corresponding stream function is

$$\psi = \frac{M\sin\alpha}{8\pi\nu t}\,\mathrm{e}^{-x}, \qquad x = \frac{r^2}{4\nu t}. \tag{3.6.19}$$

Adding the terms thus obtained to the solution (3.5.25) and (3.5.26), we find the solution for an impulsive quadrupole possessing angular momentum $N = M\sin\alpha$:

$$\omega = -\frac{M}{8\pi\nu^2 t^2}[x\mathrm{e}^{-x}\sin(2\theta-\alpha) - (1-x)\,\mathrm{e}^{-x}\sin\alpha], \tag{3.6.20}$$

$$\psi = -\frac{M}{8\pi\nu t}\{[x^{-1} - (1+x)\,\mathrm{e}^{-x}]\sin(2\theta-\alpha) - \mathrm{e}^{-x}\sin\alpha\}. \tag{3.6.21}$$

When the source acts continuously, the intensity of the source of angular momentum is $Q\sin\alpha$. The dimensions of Q are L^4T^{-2}, and hence $m = 0$ in (3.5.6). The general solution of the resulting equation is given by (3.5.7). The second term in (3.5.7) does not conserve angular momentum, since the integral (3.6.11) for the vorticity given by this term becomes infinitely large a $x \to \infty$. Thus we have $C_2 = 0$ in (3.5.7). Substituting the first term of (3.5.7) into (3.6.11), we find $C_1 = -\frac{1}{2}$, which gives

$$\omega = -\frac{Q\sin\alpha}{8\pi\nu^2 t}\,\mathrm{e}^{-x},$$

$$\psi = \frac{Q\sin\alpha}{8\pi\nu}\int_x^\infty \frac{\mathrm{e}^{-y}}{y}\,\mathrm{d}y.$$

Adding the terms thus obtained to (3.5.27) and (3.5.28), we finally obtain the solution

$$\omega = -\frac{Q}{8\pi\nu^2 t}[(1+x^{-1})\,\mathrm{e}^{-x}\sin(2\theta-\alpha) + \mathrm{e}^{-x}\sin\alpha], \tag{3.6.22}$$

$$\psi = -\frac{Q}{8\pi\nu}\left[x^{-1}(1-e^{-x})\sin(2\theta-\alpha) - \sin\alpha\int_x^\infty \frac{e^{-y}}{y}dy\right]. \quad (3.6.23)$$

The angular momentum of this quadrupolar flow is equal to $Qt\sin\alpha$ and increases linearly with time.

The additional terms in (3.6.21) and (3.6.23) give an additional azimuthal velocity that does not depend on the polar angle θ:

$$v = -\frac{\partial\psi}{\partial r} = \frac{M\sin\alpha}{4\pi\nu tr}xe^{-x}$$

for impulsive forcing and

$$v = \frac{Q\sin\alpha}{4\pi\nu r}e^{-x}$$

for continuous forcing.

Note that the linear solutions obtained can be found formally with the help of Figure 3.11 by means of the superposition (and taking the limit $\varepsilon\to 0$) of two solutions (3.5.13) or (3.5.15) for two proper dipoles, which in turn can be constructed by the superposition of two proper solutions for two point vortices. The solution for the impulsive point vortex ($\Gamma = \text{const}$), given by (3.5.10), is used to construct the impulsive dipole, while the point vortex of increasing intensity with $\Gamma = Gt$ ($G = \text{const}$), described by the stream function

$$\psi = \frac{Gt}{4\pi}\left[e^{-x} - (1+x)\int_x^\infty \frac{e^{-y}}{y}dy - \log x\right], \quad (3.6.24)$$

is used to construct the dipole of increasing intensity.

3.6.3 *Steady quadrupolar flow*

The remarkable feature of the flow described by the stream function (3.6.23) is that it becomes steady in the limit as $t\to\infty$. Using the fact that

$$\int_x^\infty \frac{e^{-y}}{y}dy \sim \log x \quad (x\to 0)$$

we obtain the steady asymptotic form of this flow:

$$\psi = -\frac{Q}{8\pi\nu}[\sin(2\theta-\alpha) + 2\sin\alpha\log r] + \psi_0(t), \tag{3.6.25}$$

$$u = \frac{1}{r}\frac{\partial\psi}{\partial\theta} = -\frac{Q}{4\pi\nu r}\cos(2\theta-\alpha), \tag{3.6.26}$$

$$v = -\frac{\partial\psi}{\partial r} = \frac{Q\sin\alpha}{4\pi\nu r}, \tag{3.6.27}$$

where $\psi_0(t) \sim -\log t$ is an inessential additional term. The pattern of particle trajectories in this flow can be obtained by solving the system

$$u = \frac{\mathrm{d}r}{\mathrm{d}t} = -\frac{Q}{4\pi\nu r}\cos(2\theta-\alpha),$$

$$\frac{v}{r} = \frac{\mathrm{d}\theta}{\mathrm{d}t} = \frac{Q\sin\alpha}{4\pi\nu r^2},$$

which gives

$$r = r_0 \exp\left[-\frac{\sin(2\theta-\alpha)}{2\sin\alpha}\right], \qquad r_0 = r(\theta_0). \tag{3.6.28}$$

The pattern varies with the variation of the angle α (shown in Figure 3.11). When $\alpha = \frac{1}{2}\pi$, the fluid particles describe closed loops elongated in the directions $\theta = 0, \pi$, where $u = 0$, and narrowed in the directions $\theta = \pm\frac{1}{2}\pi$, where u is also zero (Figure 3.16). The loops oriented in the directions where $u = 0$ become more elongated as α decreases (Figure 3.16b). When α is small, symmetrical regions with almost radial inflow and outflow form, so that the trajectories close up very far from the origin (Figure 3.16c). For a pure force dipole ($\alpha = 0$) the flow becomes completely radial and the trajectories close up at infinity.

3.7 Nonlinear multipoles

The viscous time-dependent nonlinear solutions are determined by the convection–diffusion equation for the vorticity (2.1.22), which

(a)
(b)
(c)

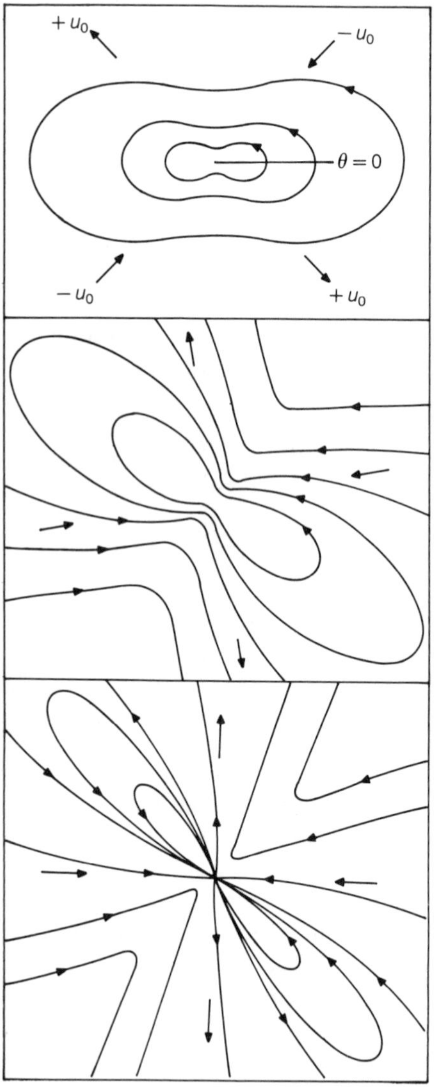

Figure 3.16 *Trajectories of fluid particles in the steady flow (3.6.26) and (3.6.27) for different values of* α *(Figure 3.11): (a)* $\alpha = 45°$*; (b) 20°; (c) 5°. The arrows indicate the maxima (* $+U_0$*) and minima (* $-U_0$*) of the radial velocity.*

can be written in the form

$$\nu \nabla^2 \omega - \frac{\partial \omega}{\partial t} = u \frac{\partial \omega}{\partial r} + \frac{1}{r} v \frac{\partial \omega}{\partial \theta}. \tag{3.7.1}$$

For the Stokes solutions the vorticity satisfies the linear part of (3.7.1), i.e. the homogeneous diffusion equation

$$\nu \nabla^2 \omega_1 - \frac{\partial \omega_1}{\partial t} = 0.$$

Here we signify by the subscript 1 the linear first-order solutions for the vortex multipoles obtained in section 3.6. Using dimensional analysis, these first-order solutions for the vorticity can be presented in the form

$$\omega_1 = Re \frac{\nu}{r^2} W_1(\eta) \sin n\theta, \qquad \eta = \frac{r}{2(\nu t)^{1/2}}, \tag{3.7.2}$$

where $Re = A\nu^p r^m$ is a non-dimensional parameter characterizing the intensity of the forcing. The constants p and $m = -(s + 2p)$ are determined by the dimensions $L^s T^p$ of the dimensional forcing amplitude A, while n is the order of the multipole. Obviously this form of solution is equivalent to (3.5.3). In the latter t, η and θ have been used as the basic arguments, while in (3.7.2) the basic arguments are r, η and θ. As with (3.7.2), the expressions for the first-order stream functions and for the velocity components can be written in the form

$$\left.\begin{aligned} \psi_1 &= Re\, \nu \Psi_1(\eta) \sin n\theta, \\ u_1 &= Re \frac{\nu}{r} U_1(\eta) \cos n\theta, \\ v_1 &= Re \frac{\nu}{r} V_1(\eta) \sin n\theta. \end{aligned}\right\} \tag{3.7.3}$$

The non-dimensional functions Ψ_1, W_1, U_1 and V_1 are given by the linear Stokes solutions obtained in the preceding sections for each multipole.

Substitution of the first-order solutions (3.7.2) and (3.7.3) into

(3.7.1) permits estimation of the nonlinear part of (3.7.1):

$$u_1 \frac{\partial \omega_1}{\partial r} + \frac{1}{r} v_1 \frac{\partial \omega_1}{\partial \theta} \propto Re^2 \sin 2n\theta .$$

It can be seen that this nonlinear term is of order Re^2 and contains the second-order angular harmonic $\sin 2n\theta$. This prompts one to consider the second-order solutions for the vorticity in the form

$$\omega_2 = Re^2 \frac{v}{r^2} W_2(\eta) \sin 2n\theta . \tag{3.7.4}$$

The second-order solution ω_2 has to satisfy the equation

$$v \nabla^2 \omega_2 - \frac{\partial \omega_2}{\partial t} = u_1 \frac{\partial \omega_1}{\partial r} + \frac{1}{r} v_1 \frac{\partial \omega_1}{\partial \theta}, \tag{3.7.5}$$

which includes the transport of vorticity by the first-order flow. Since the right-hand side of (3.7.5) contains the known first-order solutions, we obtain the linear equation for the second-order approximation.

The linear solutions satisfy the equation of motion in the limit as $Re \to 0$; hence the sum $\omega = \omega_1 + \omega_2$ will be a better approximation for $Re \ll 1$. The second-order stream functions can be found from the Poisson equation with the second-order vorticity

$$\nabla^2 \psi_2 = -\omega_2 . \tag{3.7.6}$$

Here we assume that ψ_2 is given by

$$\psi_2 = Re^2 v \Psi_2(\eta) \sin 2n\theta .$$

Thus we have to find the second term in the asymptotic expansion

$$\omega = \frac{v\, Re}{r^2} [W_1 + Re\, W_2 + O(Re^2)],$$

where Re is the small parameter. Since Re gives the order of magnitude of the ratio of the convective (nonlinear) term to the viscous (linear) term in the equation of motion, it can be considered as the Reynolds number of the flow.

The question arises as to whether one can continue this asymptotic expansion to higher orders? This is not, obviously, a matter of routine, since only the second-order term ω_2 permits one to determine the angular dependence in the simple form $\omega_2 \propto \sin 2n\theta$. The nonlinear terms in the equation of motion cause the creation of an increasing number of different angular harmonics, and the problem for higher-order approximations becomes too complicated for analytical treatment.

Before considering some particular nonlinear vortex multipoles, it is useful to transform the partial differential equation (3.7.5) into an ordinary differential equation. Substitution of ω_1, u_1, v_1 and ω_2 in the form (3.7.2), (3.7.3) and (3.7.4) into (3.7.5) gives the general governing equation for the second-order non-dimensional vorticity W_2 in the form

$$\eta^2\frac{\mathrm{d}^2W_2}{\mathrm{d}\eta^2}+[2\eta^3+(4m-3)\eta]\frac{\mathrm{d}W_2}{\mathrm{d}\eta}+[(2m-2)^2-(2n)^2]W_2$$
$$=\frac{1}{2}\left[(m-2)U_1W_1+\eta U_1\frac{\mathrm{d}W_1}{\mathrm{d}\eta}+nV_1W_1\right], \qquad (3.7.7)$$

where m and n are the numbers characterizing the particular forcing and the regime of its action. Similarly, one can obtain from (3.7.6) the general equation for the second-order stream function Ψ_2 in the form

$$\eta^2\frac{\mathrm{d}^2\Psi_2}{\mathrm{d}\eta^2}+(4m+1)\eta\frac{\mathrm{d}\Psi_2}{\mathrm{d}\eta}+4(m^2-n^2)\Psi_2=-W_2. \qquad (3.7.8)$$

3.7.1 *Nonlinear dipole*

Consider, for example, a point momentum source acting continuously: it starts at time $t=0$ and thereafter exerts on the fluid a force ρJ in the direction $\theta=0$. The dimensions of the main governing parameter $A=J$ are L^3T^{-2}, which gives $p=-2$ and $m=1$. Since for a dipole $n=1$, (3.7.7) becomes

$$\eta^2\frac{\mathrm{d}^2W_2}{\mathrm{d}\eta^2}+(2\eta^3+\eta)\frac{\mathrm{d}W_2}{\mathrm{d}\eta}-4W_2=\frac{1}{2}\left(-U_1W_1+\eta U_1\frac{\mathrm{d}W_1}{\mathrm{d}\eta}+V_1W_1\right). \qquad (3.7.9)$$

Normalizing for the sake of simplicity the Reynolds number Re in

(3.7.2) by 2π,

$$Re = \frac{Jr}{2\pi\nu^2},$$

we find from the linear solution (3.5.14) and (3.5.15), the first-order functions W_1, U_1 and V_1 in the form

$$W_1 = e^{-\eta^2},$$

$$\left.\begin{aligned} U_1 &= \frac{1}{4}\left[\eta^{-2}(1-e^{-\eta^2}) + \int_{\eta^2}^{\infty} e^{-x}x^{-1}\,dx\right], \\ V_1 &= \frac{1}{4}\left[\eta^{-2}(1-e^{-\eta^2}) - \int_{\eta^2}^{\infty} e^{-x}x^{-1}\,dx\right]. \end{aligned}\right\} \quad (3.7.10)$$

This gives for the right-hand side of (3.7.9)

$$\Phi = -\frac{1}{4}e^{-\eta^2}\left[(1+\eta^2)\int_{\eta^2}^{\infty} e^{-x}x^{-1}\,dx + 1 - e^{-\eta^2}\right]. \quad (3.7.11)$$

Two homogeneous solutions of (3.7.9) are $1-\eta^{-2}$ and $\eta^{-2}e^{-\eta^2}$, and with the help of these solutions the inhomogeneous solution is obtained, using standard methods. Thus we obtain the general solution of (3.7.9) in the form

$$W_2 = C_1(1-\eta^{-2}) + C_2\eta^{-2}e^{-\eta^2} - \tfrac{1}{2}(1-\eta^{-2})\int_{\eta}^{\infty}\frac{\Phi\,d\eta}{\eta^3} - \tfrac{1}{2}\eta^{-2}e^{-\eta^2}\int_{\eta}^{\infty}\frac{\Phi e^{\eta^2}(1-\eta^2)}{\eta^3}\,d\eta,$$

where Φ is given by (3.7.11). After lengthy calculations of the integrals, the above expression gives

$$W_2 = C_1(1-\eta^{-2}) + C_2\eta^{-2}e^{-\eta^2} + \frac{1}{16}\left[\frac{e^{-\eta^2}}{\eta^2}(1-3e^{-\eta^2}+\log\eta^2) + \left(\frac{4e^{-\eta^2}}{\eta^2} + e^{-\eta^2} + \frac{1}{\eta^2} - 1\right)\int_{\eta^2}^{\infty} e^{-y}y^{-1}\,dy + \left(4 - \frac{4}{\eta^2}\right)\int_{2\eta^2}^{\infty} e^{-y}y^{-1}\,dy\right]. \quad (3.7.12)$$

Here the first two terms represent the general solution of the homogeneous equation, and contain two constants C_1 and C_2, which have to be determined with the help of appropriate boundary conditions at $\eta = 0$ and as $\eta \to \infty$. The choice $C_1 = 0$ ensures that $W_2(\infty) = 0$. To find C_2, we must analyse the leading terms of (3.7.12) as $\eta \to 0$.

The integrals in (3.7.12) are both particular values of the well-known exponential integral

$$E_1(x) = \int_x^\infty e^{-y} y^{-1}\, dy.$$

This can be expanded for $x \to 0$ as follows (e.g. Korn and Korn, 1961; Arfken 1985):

$$E_1(x) = -\left(\gamma + \log x - \frac{x}{1!1} + \frac{x^2}{2!2} - \frac{x^3}{3!3} + \cdots\right),$$

where $\gamma = 0.5772\ldots$ is the Euler–Mascheroni constant. Using this expansion, we obtain the coefficient of the leading term η^{-2} of (3.7.12) in the form $[C_2 - \frac{1}{16}(2 - 4\log 2 + \gamma)]\eta^{-2}$. Requiring the minimum possible singularity at the origin ($\eta = 0$), we immediately obtain the value of the second constant: $C_2 = \frac{1}{16}(2 - 4\log 2 + \gamma)$. In this case the leading term of (3.7.12) is $\log \eta$. Thus the second-order solution for the vortex dipole when the force acts continuously is given by

$$\omega = \omega_1 + \omega_2 = Re \frac{v}{r^2} W_1(\eta) \sin\theta + Re^2 \frac{v}{r^2} W_2(\eta) \sin 2\theta, \qquad (3.7.13)$$

where $Re = (2\pi)^{-1} J v^{-2} r$, and W_1 and W_2 are given by (3.7.10) and (3.7.12) respectively. (For an impulsive dipole the second-order solution can be found in Cantwell and Rott (1988)). One could also obtain the second-order stream function of the flow, substituting W_2 into (3.7.8) and solving the resulting equation. However, the calculations are too lengthy to be given here.

Several features are particularly noteworthy in the results obtained. At the origin ($r \to 0$), where the force acts continuously, the first-order term in (3.7.13) is algebraically singular, $\omega_1 \propto r^{-1}$, while the second-order term is less so, $\omega_1 \propto \log r$. Thus the requirement that higher-order approximations be no more singular than the first-order solution (Van Dyke, 1975) is satisfied. On the other hand, although the first-order solution for the vorticity gives steady asymptotic behaviour, since $\omega_1 \propto r^{-1}$ as $t \to \infty$, the first-order solutions for the velocity components (3.7.3) and (3.7.10) do not, because of the presence of the logarithmic terms $u_1, v_1 \propto \log t$, which tend to infinity as $t \to \infty$. Hence the radius of validity of the first-order solution tends to zero as $t \to \infty$.

It turns out that the second-order solution for the vorticity does not give steady asymptotic behaviour either, since $\omega_2 \propto \log t$ as $t \to \infty$. Thus the question raised by Cantwell (1986) as to whether or not an exact nonlinear solution exists for the planar steady jet remains unsolved. It is known, however, that there is steady asymptotic behaviour for the appropriate three-dimensional axisymmetric flow induced by a point force of intensity ρJ. In this case the dimensions of J are L^4T^{-2}. This asymptotic behaviour is given by the well-known solution for the steady round jet obtained by Slezkin (1934) and Landau (1944) and studied in detail by Squire (1951). The appropriate exact steady solution for the planar jet has still not been found, only Schlichting's (1933) solution obtained in the boundary layer approximation is known.

In spite of the above-mentioned shortcomings, the weakly nonlinear solution given by (3.7.13) appears to be very useful. In particular, this solution allows the vortex dipole to drift forward, along the axis $\theta = 0$, while the linear solution describes only the homogeneous expansion of the vortex structure relative to the point $r = 0$ (Figure 3.9a). Thus the nonlinearity causes translational motion of the dipole.

3.7.2 *Nonlinear quadrupole*

Consider now a point force dipole ($\alpha = 0$) acting continuously: it starts at time $t = 0$ and thereafter acts on the fluid with the intensity Q in the directions $\theta = \pi, 0$ (Figure 3.11). The dimensions of Q are L^4T^{-2}, which gives $p = -2$ and $m = 0$ for Re in (3.7.2). Normalizing for the sake of simplicity the Reynolds number by 4π, $Re = Q/4\pi\nu^2$, we obtain from the linear solution (3.6.22) and (3.6.23) the first-order functions W_1, U_1 and V_1 in (3.7.2) and (3.7.3) in the form

$$\left.\begin{aligned} W_1 &= -2(\eta^2+1)e^{-\eta^2}, \\ U_1 &= -\eta^2(1-e^{-\eta^2}), \\ V_1 &= -\eta^{-2}+(1+\eta^{-2})e^{-\eta^2}. \end{aligned}\right\} \tag{3.7.14}$$

Substituting (3.7.14) into the right-hand side of (3.7.7) gives for $m = 0$

and $n = 2$ the governing equation in the form

$$\eta^2 \frac{d^2 W_2}{d\eta^2} + (2\eta^3 - 3\eta)\frac{dW_2}{d\eta} - 12 W_2 = -2(\eta^2 + e^{-\eta^2})e^{-\eta^2}.$$

With the help of the new variable $x = \eta^2$, the above equation becomes

$$x^2 \frac{d^2 W_2}{dx^2} + x(x-1)\frac{dW_2}{dx} - 3 W_2 = -\tfrac{1}{2}(x + e^{-x})e^{-x}. \qquad (3.7.15)$$

Two homogeneous solutions of (3.7.15) are $1 - 3x^{-1}$ and $(x + 4 + 6x^{-1})e^{-x}$. Then, following the standard routine and writing the inhomogeneous solution of (3.7.15) in integral form, we obtain the general solution of (3.7.15) in the form

$$\begin{aligned} W_2 = {} & C_1(1 - 3x^{-1}) + C_2(x + 4 + 6x^{-1})e^{-x} \\ & - (1 - 3x^{-1}) \int_x^\infty \frac{\Phi}{x^3}(x + 4 + 6x^{-1})\,dx \\ & - e^{-x}(x + 4 + 6x^{-1}) \int_x^\infty \frac{\Phi e^x}{x^3}(3x^{-1} - 1)\,dx, \end{aligned} \qquad (3.7.16)$$

where Φ is the right-hand side of (3.7.15). After lengthy calculations of the two integrals in (3.7.16), we obtain the general solution of (3.7.15) in the form

$$\begin{aligned} W_2 = {} & C_1(1 - 3x^{-1}) + C_2(x + 4 + 6x^{-1}) - \tfrac{1}{2}\Big[(\tfrac{3}{2}x^{-1} + 1)e^{-x} \\ & - (1 + 4x^{-1})e^{-2x} + (x + 4 + 6x^{-1})e^{-x} \int_x^\infty e^{-y} y^{-1}\,dy \\ & + (2 - 6x^{-1}) \int_x^\infty e^{-y} y^{-1}\,dy\Big]. \end{aligned} \qquad (3.7.17)$$

Requiring that W_2 decay at infinity ($x \to \infty$), we immediately have $C_1 = 0$. To find the second constant C_2, we must analyse the behaviour of the function W_2 at the origin ($x = 0$). Expanding (3.7.17), we obtain

$$\lim_{x \to 0} W_2(x) \approx C_2(2 - 6x^{-1}) - \tfrac{1}{2}[(\tfrac{1}{2} - 2\log 2) + (6\log 2 - \tfrac{5}{2})x^{-1}]. \qquad (3.7.18)$$

The coefficient of the leading term x^{-1} in (3.7.18) is $-6C_2 -$

$3\log 2+\frac{5}{4}$. We now recall that the first-order solution ω_1 is regular at $r=0$ $(x=0)$. From general principles we also require regularity for the second-order solution ω_2 at the point $r=0$. This gives for C_2 in (3.7.17) the value $C_2=\frac{5}{24}-\frac{1}{2}\log 2$ and $\lim_{x\to 0} W_2(x)=\frac{1}{6}$. The resulting expression for W_2 is

$$W_2=[(\tfrac{1}{2}\log 2-\tfrac{5}{24})x+(2\log 2-\tfrac{4}{3})+(3\log 2-2)x^{-1}]e^{-x}$$
$$+(2x^{-1}+\tfrac{1}{2})e^{-2x}-(3x^{-1}+2+\tfrac{1}{2}x)e^{-x}\int_x^{\infty} e^{-y}y^{-1}\,dy$$
$$-(1-3x^{-1})\int_x^{\infty} e^{-2y}y^{-1}\,dy. \qquad (3.7.19)$$

Thus the second-order solution for the quadrupolar flow induced by a point force dipole acting continuously is

$$\omega=\omega_1+\omega_2=Re\frac{v}{r^2}W_1(\eta)\sin 2\theta+Re^2\frac{v}{r^2}W_2(\eta)\sin 4\theta, \qquad (3.7.20)$$

where $Re=Q/4\pi v^2$, and W_1 and W_2 are given by (3.7.14) and (3.7.19) respectively.

In contrast to the solution (3.7.13) for the dipole, it turns out that the second-order solution (3.7.20) for the quadrupole is finite in the limit as $t\to\infty$. Moreover, all properties of the quadrupolar flow have a steady asymptotic behaviour, and in the limit as $t\to\infty$ they are given by

$$\left.\begin{aligned}
\omega&=-2\,Re\frac{v}{r^2}\sin 2\theta+\tfrac{1}{6}Re^2\frac{v}{r^2}\sin 4\theta,\\
\psi&=-\tfrac{1}{2}Re\,v\sin 2\theta+\tfrac{1}{96}Re^2\,v\sin 4\theta,\\
u&=-\,Re\frac{v}{r}\cos 2\theta+\tfrac{1}{24}Re^2\frac{v}{r}\cos 4\theta,\\
v&=0.
\end{aligned}\right\} \qquad (3.7.21)$$

These expressions describe a steady flow consisting of two jets flowing in the directions $\theta=\pm\frac{1}{2}\pi$ from the origin and two inflow regions with axes $\theta=0,\pi$ where the fluid moves towards the origin (Figure 3.17). Since the azimuthal velocity is zero, the fluid particles move along straight radial lines, in both the inflow and outflow

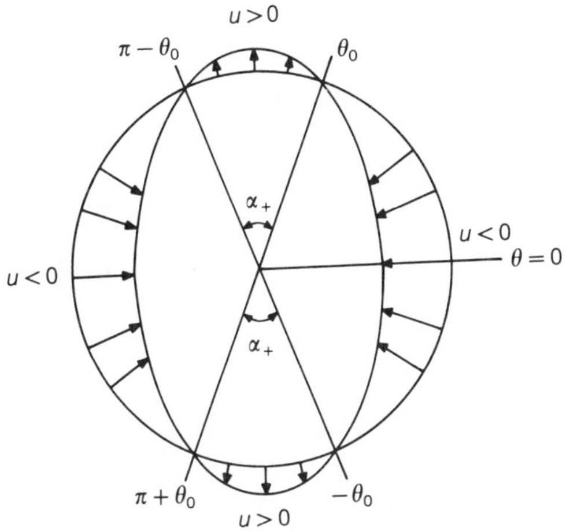

Figure 3.17 *Schematic drawing of a steady quadrupolar flow, showing outflow sectors and inflow sectors. The fluid particles move along the radial lines* $\theta = const$. *The apex angles* ($\alpha_+ = \pi - 2\theta_0$) *of the outflow sectors decrease when the intensity of forcing increases.*

regions. The radial velocity vanishes at the lines $\theta = \pm\theta_0$ and $\theta = \pi \pm \theta_0$, where the value of the angle θ_0 is determined by the relation $\cos 2\theta_0 - \frac{1}{24} Re \cos 4\theta_0 = 0$.

The outflow is distributed over two sectors with apex angles $\alpha_+ = \pi - 2\theta_0$, which are less than that of the inflow sectors, $\alpha_- = 2\theta_0$, since $\theta_0 = \frac{1}{4}\pi$ when $Re \to 0$ and α_+ decreases when Re increases. Thus the outflow jets become narrower as the intensity of the forcing grows.

To reproduce such a flow experimentally, a vertical rod oscillating horizontally in a thin layer of fluid is used (section 5.4). Disregarding the small region in the vicinity of the oscillating rod where the motion is essentially unsteady, it can be concluded that the real flow picture (Figure 5.17c) agrees well with the description given with the help of the steady solution (3.7.21).

To obtain the second-order solution for a rotating quadrupole ($\alpha \neq 0$), which possesses angular momentum, one must solve (3.7.5) for ω_1 given by (3.6.22). The part of ω_1 in (3.6.22) independent of θ gives an additional term $W_2^* \propto \sin(2\theta - \alpha)$ in the second-order solution (3.7.20).

4

Vortex dipole interactions in a stratified fluid

Various irregular quasi-two-dimensional flows associated with the term 'two-dimensional turbulence' are induced by a system of forces applied to a fluid. The action of a localized forcing plus gravity leads to the formation of coherent vortices in a stratified fluid. It seems that vortex dipoles and their combinations are the universal products of any irregular forcing in two-dimensional fluid systems. The vortex dipole is characterized by non-zero linear momentum and can be considered as one of the basic ordered structures in two-dimensional flows. In view of its interaction properties, it has been termed the 'elementary particle' of two-dimensional chaotic flows. Vortex dipole interactions of different kinds lead to the emergence of more complex ordered structures. Thus it seems reasonable to reproduce and study separately basic types of dipole interactions. A simple linear mathematical model of the symmetric collision of two dipoles (Figure 3.14) permits one to analyse the dynamics of the interaction as well as to obtain asymptotically the quadrupolar flow. It turns out that, in spite of their linear character, the analytical solutions can be very useful in the interpretation of real interactions. Moreover, the vortex multipoles that were considered as a result of the expansion (2.1.28) determine the asymptotics of a real flow that is initially turbulent. The analytical results can help to reveal the implicit order in a very complicated flow and to predict the ultimate stage of flow evolution. In this chapter, following the papers by Voropayev and Afanasyev (1992, 1993a), we consider experimental results on vortex dipole interactions in a stratified fluid and interpret these in terms of forces acting on the fluid and associated vortex multipoles: dipoles, quadrupoles and their combinations.

4.1 Symmetric collision of two dipoles of equal momenta

Consider two sources of momentum that start simultaneously and thereafter apply to the fluid equal forces, acting in opposite directions for a short time interval Δt. After the onset, the sources generate two initially turbulent strong jets ($Re \approx 700$, Figure 4.1a). The jets, which have equal momenta ($I_1 = I_2 = I$), collide, forming a chaotic turbulent three-dimensional cloud (Figure 4.1b). Owing to the effect of gravity, the flow becomes quasi-planar. As a result of the collision, two dipoles are pushed from the cloud in opposite directions, perpendicular to the initial directions of motion of the jets (Figures 4.1c, d).

Consider now the same interaction but for laminar flows ($Re \approx 45$, Figure 4.2). The initial jets are weak and the flow is laminar from the start; thus the jets form two dipoles before the collision (Figure 4.2a). These dipoles move towards one another, forming a quadrupole (Figure 4.2b) consisting of the vortices of the primary dipoles. Two newly formed dipoles then move apart perpendicular to the initial directions of motion of the primary dipoles (Figures 4.2c, d). Although the initial flows in Figures 4.1(a) and 4.2(a) look very different, the final result of the interactions (Figures 4.1d, 4.2d) – the formation of dipoles of equal intensity moving apart – is the same for both cases (see also Figure 3.14). It is clear that, at least qualitatively, we can consider the interaction dynamics without emphasizing the regime of the flow – whether it is initially turbulent or laminar – and below we present only the most illustrative examples.

Now consider the same interaction in terms of forcing and associated vortex multipoles. In this case the total momentum of the flow is zero; hence, there is only one parameter: the quadrupole intensity of the flow $M = I\varepsilon$, where ε is the distance between the two sources. After the sources start, two primary dipoles form. Initially, when $R < \frac{1}{2}\varepsilon$ (where R is a typical radius of the dipole), they develop independently, governed by the parameters I_1 and I_2 ($I_1 = I_2 = I$) respectively (Figure 4.2a). The vorticity generated by the sources is concentrated approximately in the area contoured by the dyed fluid. When the dipole radius R, which is also the length scale of the vortex patches, becomes comparable to the distance between the two patches, the interaction begins. The mutual induction of two dipolar vortex patches leads to the emergence of potential quadrupolar backflow outside the patches (Figure 4.3). At the moment of collision, vorticity

(a)

(b)

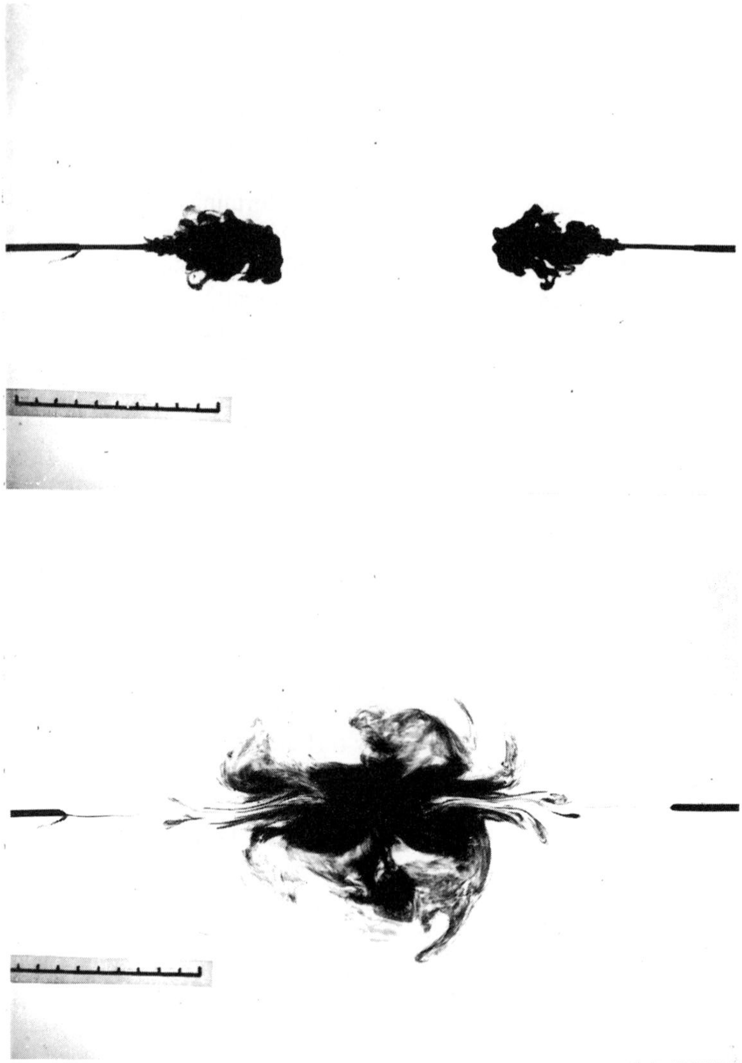

Figure 4.1 *Photographic sequence showing a top view of the collision of two impulsive strong jets ($Re \approx 700$) of equal momentum and the formation of an impulsive vortex quadrupole in a linearly stratified ($\mathcal{N} = 1.5\,s^{-1}$) fluid. The scale is in centimetres. The distance between sources $\varepsilon = 15$ cm.*

(c)

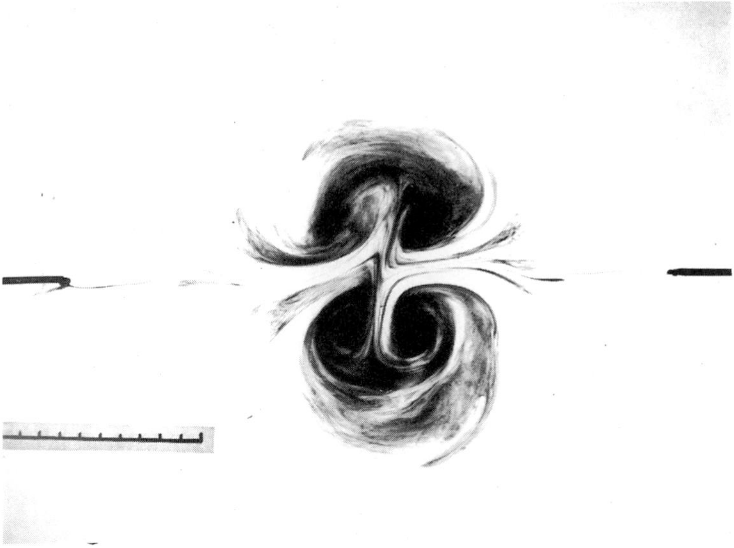

(d)

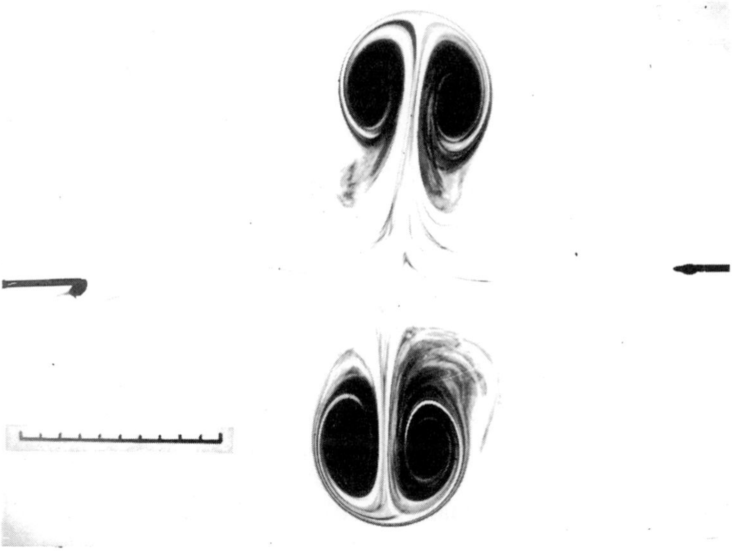

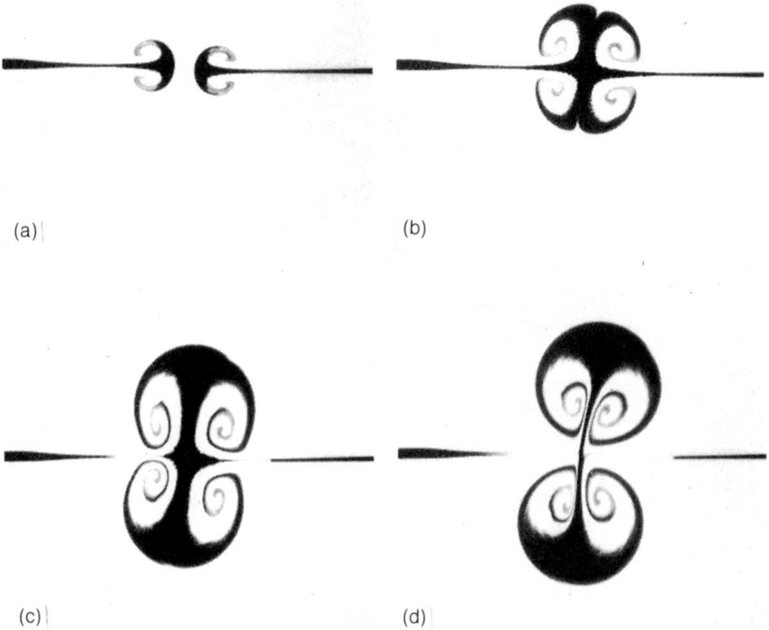

Figure 4.2 *Photographic sequence showing the symmetric collision of two dipoles ($Re \approx 45$) of equal momentum and the formation of a vortex quadrupole. The distance between sources $\varepsilon = 2.5$ cm, $\mathcal{N} = 1.5\ s^{-1}$, and the time after the start of the experiment $t = 2$ s (a), 7 s (b), 13 s (c) and 19 s (d).*

is localized around the collision point (Figure 4.2b). Thus a localized vortex quadrupole forms in this region (Figure 4.3b). The subsequent flow evolution is governed by the single parameter M. As a consequence, the quadrupolar flow develops in accordance with the theoretical solution (3.5.26) for the point source shown in Figure 3.12.

More realistic dyed water distributions for the interaction shown in Figure 4.2 can be obtained by the superposition of two solutions (3.5.15) for two equal-amplitude point sources of momentum ($\varepsilon \neq 0$). Calculating the movement of marked particles in the resulting flow, we obtain the pictures shown in Figure 3.14, which give a good

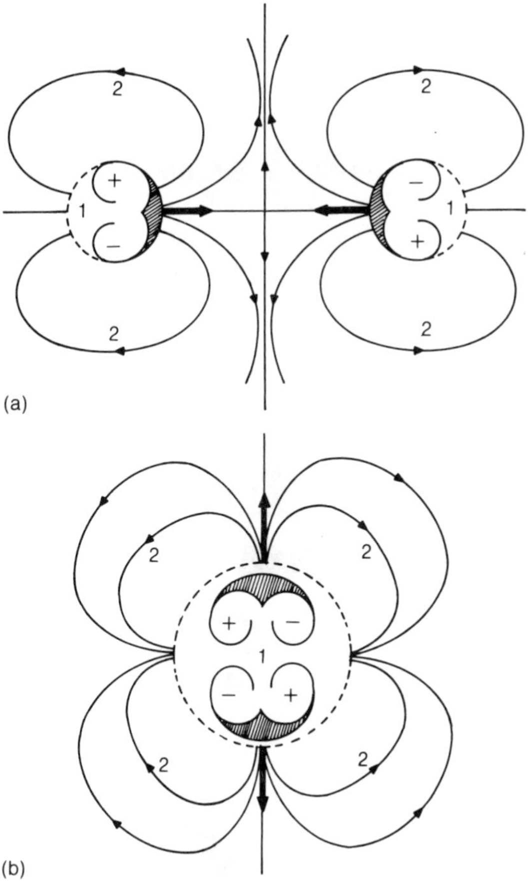

Figure 4.3 *Sketch of the symmetric collision of two dipoles of equal momentum: (a) initial phase of interaction; (b) formation of compact vortex quadrupole; 1 = vortical flow; 2 = potential flow.*

qualitative approximation of the real flow evolution shown in Figure 4.2.

4.2 Head-on collision of two dipoles of different momenta

Consider the case when two colliding jets have different momenta. In Figure 4.4(a) the left-hand jet acts for three times as long as the

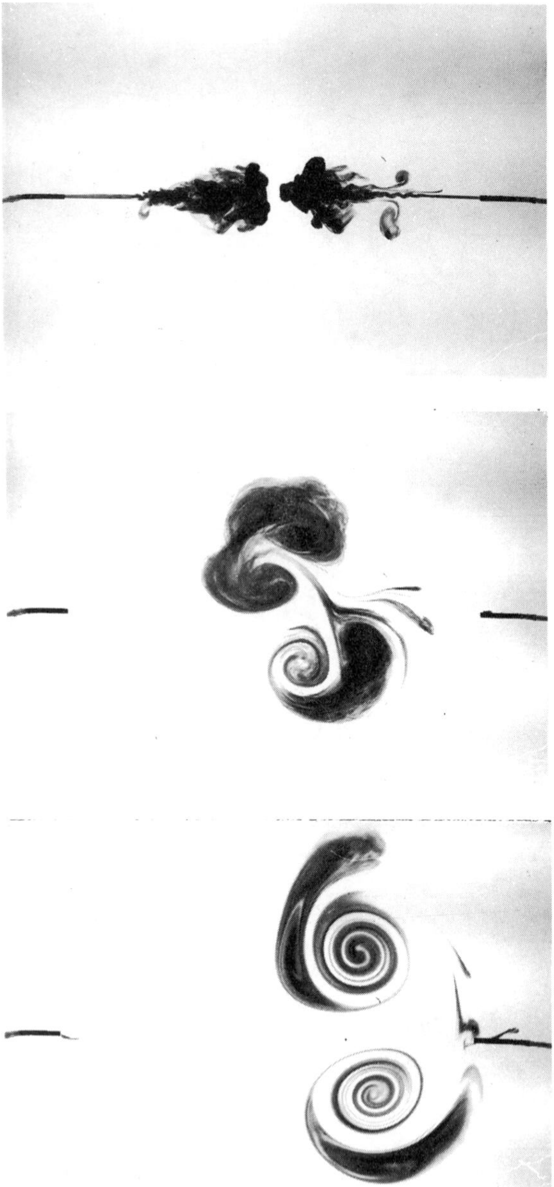

Figure 4.4 *Photographic sequence showing the collision of two impulsive strong jets (Re ≈ 700) of different momenta: two sources of equal intensity*

(Continued on next page)

(b)

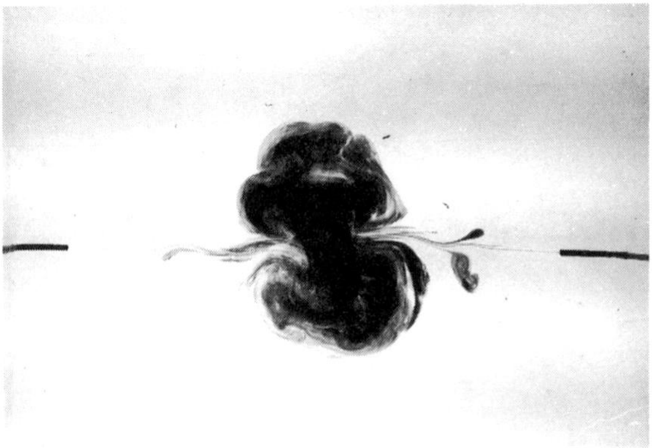

(d)

(f)

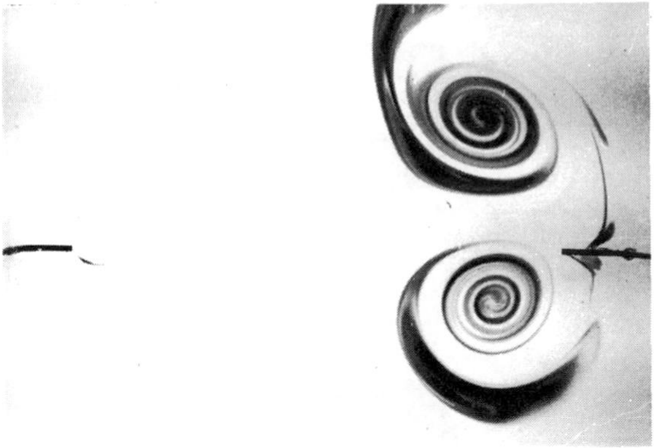

act for periods of time 6 s and 2 s respectively; thus the left-hand source applies to the fluid a total momentum three times greater than that applied by the right-hand source. The distance between sources ε = 14.5 cm.

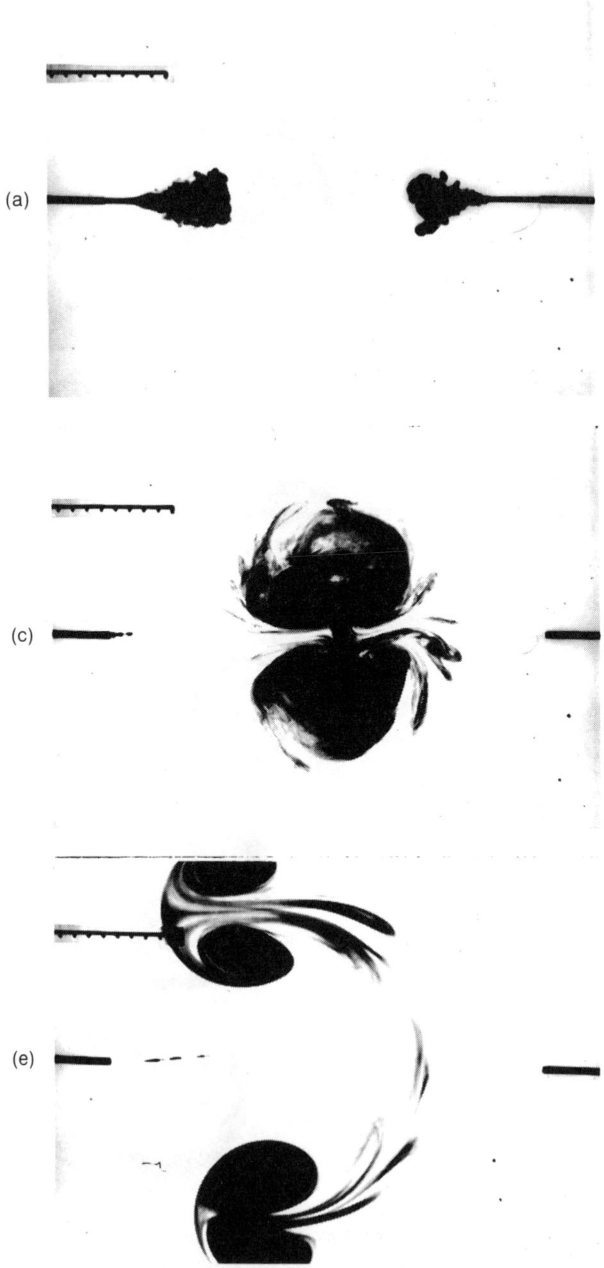

Figure 4.5 *Sequence of photographs showing the collision of two impulsive jets (*$\mathrm{Re} \approx 800$*) of different momenta: two sources of equal intensity act for*

(*Continued on next page*)

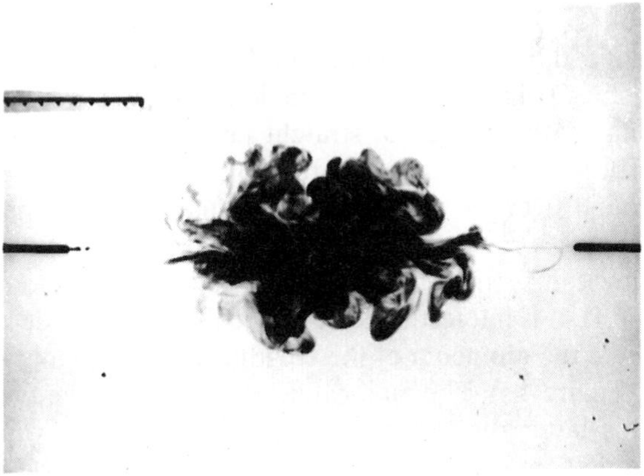

(b)

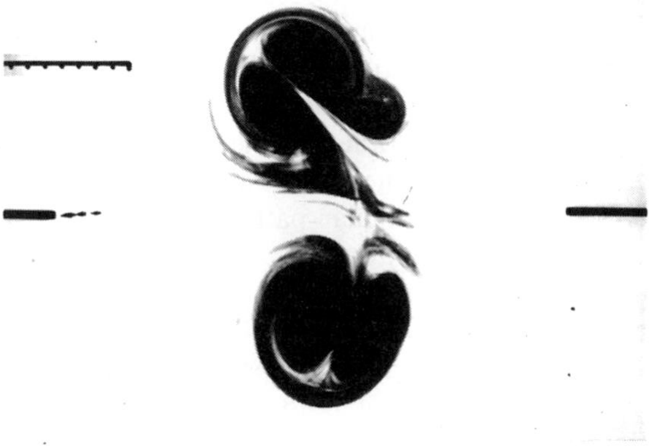

(d)

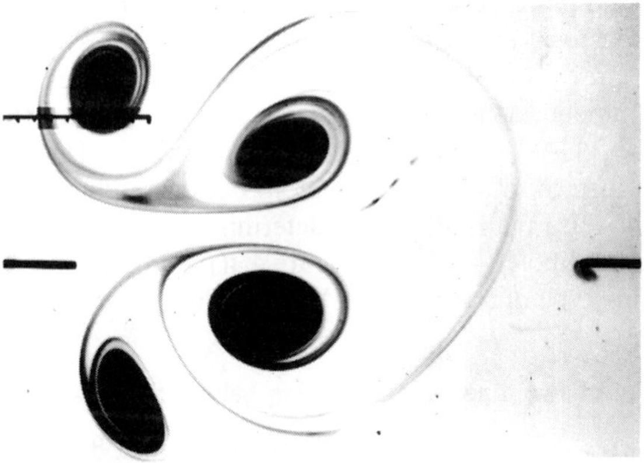

(f)

different periods of time; thus the left-hand source applies to the fluid a total momentum 1.3 times greater than that applied by the right-hand source.

right-hand one; thus $I_1 = 3I_2$. The secondary dipoles emerging from the chaotic cloud (Figures 4.4b, c) are no longer symmetric, and move along circular paths instead of straight lines (Figures 4.4c–e). The weaker vortices in these dipoles are subjected to straining while orbiting around stronger vortices, and eventually decay. Finally, the stronger vortices form a new dipole, which moves to the right, in the same direction as the primary jet that has the largest momentum (Figures 4.4e, f). It is interesting to compare the case considered with the case where the momenta of the colliding jets differ only slightly ($I_1 \approx 1.3I_2$, Figure 4.5). The major difference between the two cases is that in the latter the radius of the circular paths in which the secondary dipoles move is much larger than the radius of the dipoles. The weaker vortices in the dipoles do not decay in the weaker background straining field. As a result, two secondary dipoles consisting of vortices of different intensity perform a looping excursion and collide again near the original collision point. Finally, after another exchange of partners between the colliding dipoles, two new dipoles are formed. The newly formed dipoles have different momenta, which correspond to the momenta of the primary jets. Note that the common radius of the circular paths in which the secondary dipoles move depends on the ratio I_1/I_2. When $I_1/I_2 \to 1$, this radius tends to infinity, and we have a 'pure' quadrupole (section 4.1).

Consider the collision shown in Figure 4.4 in terms of its governing parameters. In this case both of the parameters $I = I_1 - I_2$ and $M = I_2\varepsilon$ are non-zero. The total flow can be considered as a superposition of dipolar (I) and quadrupolar (M) flows. The velocities induced by a dipole and a quadrupole decrease with distance as r^{-2} and r^{-3} respectively (section 3.5). Thus at small distances the quadrupolar flow prevails. As a result, two secondary dipoles governed by the parameter M emerge from the cloud after the collision (Figures 4.4b, c). At larger distances dipolar flow prevails. As a result, the newly formed dipoles move in a background dipolar flow (Figures 4.4c–e). Finally, the flow determined by the multipole of lowest order (dipole) survives (Figure 4.4f). (For a mathematical model of this type of collision see Figure 4.20.)

4.3 Merging of two dipoles, one moving behind the other

Consider two sources acting impulsively in the same direction along the line connecting the sources, or equivalently, one source that acts

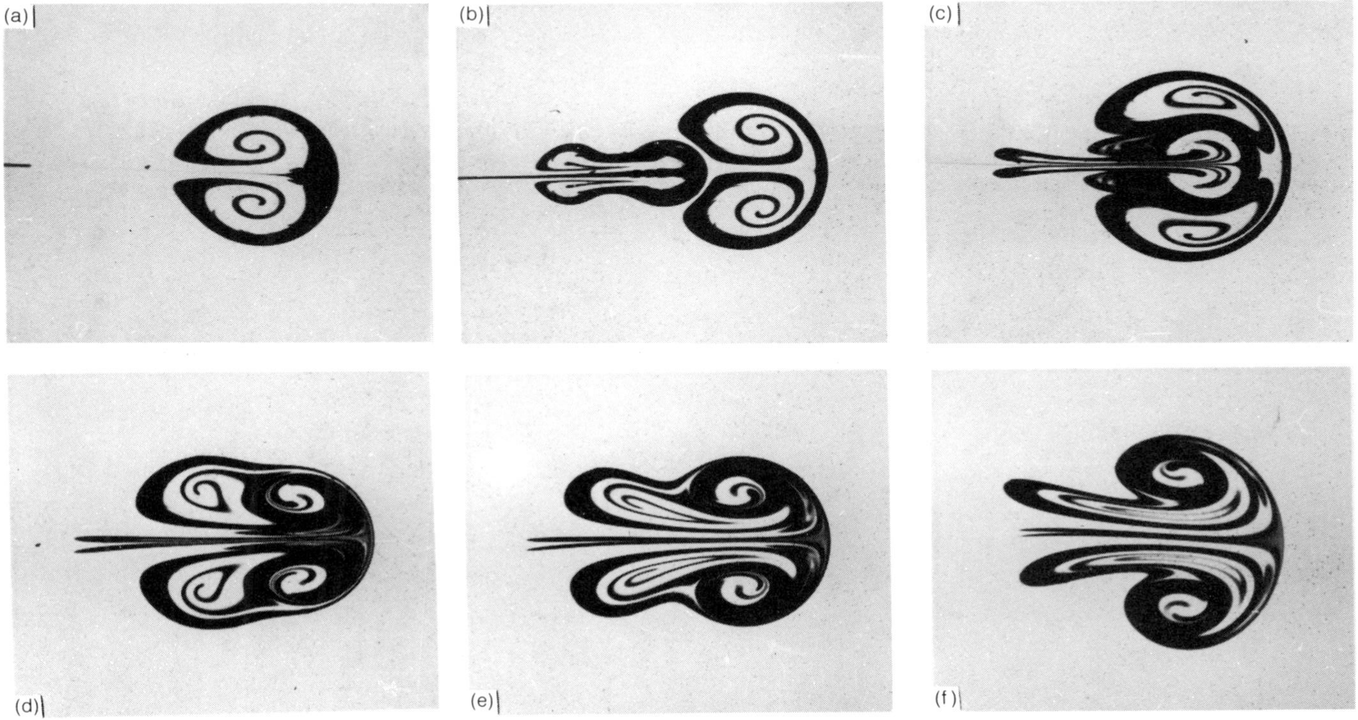

Figure 4.6 *Photographic sequence showing the interaction of two dipoles ($Re \approx 80$) moving one behind the other.*

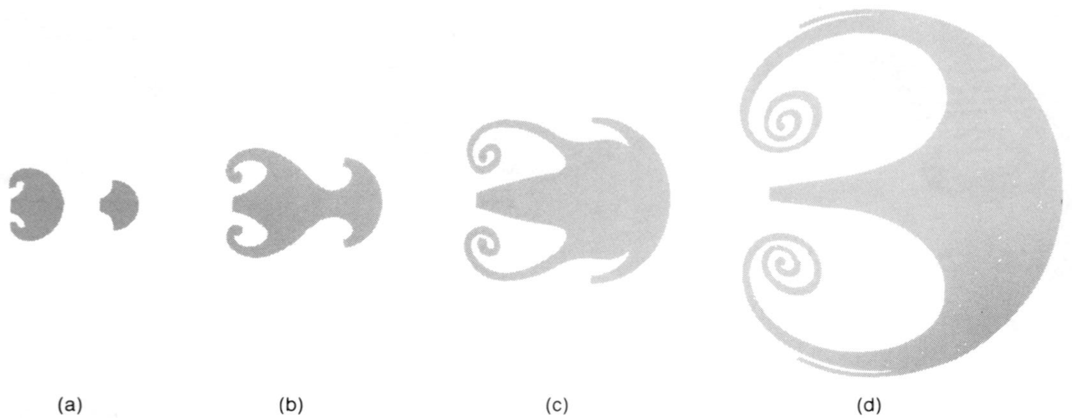

Figure 4.7 *Dyed water distributions (shaded area) for two dipoles moving one behind the other. For calculations a superposition of two solutions (3.5.15) for two momentum sources acting in the direction of the line connecting the sources was used:* $J_1 = 0.03\ cm^3\ s^{-2}$ *(the right-hand source),* $J_2 = 0.08\ cm^3\ s^{-2}$*, the distance between sources* $\varepsilon = 0.5$ *cm, and* $t = 1$ *s (a), 2 s (b), 4 s (c) and 8 s (d).*

twice. The photographic sequence in Figure 4.6 shows the flow evolution in this case. The second dipole can be seen to penetrate the first from the rear, pushing aside its vortices (Figures 4.6b, c). The strained outer vortices orbit around the vortices of the intruding second dipole (Figures 4.6c–e). This continues until a single dipole forms as a result of vortex merging (Figure 4.6f).

The explanation in this case is simple. Initially, the dipoles develop independently, governed by the parameters I_1 and I_2 respectively. The mutual induction of two dipolar vortex patches generates a net dipolar flow of intensity $I = I_1 + I_2$. The vortex patches (dyed fluid) move in this background dipolar flow; the two dipoles merge and finally form one dipole of intensity I. The results of calculations with appropriate superposition of two solutions (3.5.15) are shown in Figure 4.7. The merging of two dipoles and subsequent formation of the resulting dipole are clearly seen in this figure.

4.4 Parallel motion of two dipoles

Consider the interaction of two dipoles of equal intensity ($I_1 = I_2 = I$) moving parallel to one another. The dipoles shown in Figure 4.8 were generated by the simultaneous action of two sources of equal intensity. In the initial period after the onset of motion the dipoles do not influence one another (Figure 4.8a). As time progresses, the dimensions of the dipoles increase, and the dipoles begin to interact. They push sideways, and their inner vortices begin to orbit around the outer ones (Figures 4.8b, c). During this process the strained, initially inner, vortices decay. The remaining vortices form a new dipole, which continues to move forward (Figure 4.8d).

The interpretation in this case is similar to that in the previous one. Initially, when $R < \frac{1}{2}\varepsilon$ (ε being the distance between the sources), two primary dipoles develop independently and are clearly seen in Figure 4.8(a). The mutual induction of dipoles creates a net dipolar flow of intensity $2I$. The dipoles moving in this flow (Figures 4.8b, c) eventually form one dipole (Figure 4.8d). The results of calculations for this type of interaction are shown in Figure 4.9.

4.5 Oblique collision of two dipoles

In the case where two dipoles of equal momenta ($I_1 = I_2 = I$) collide at some angle, each of the primary dipoles splits into two vortices,

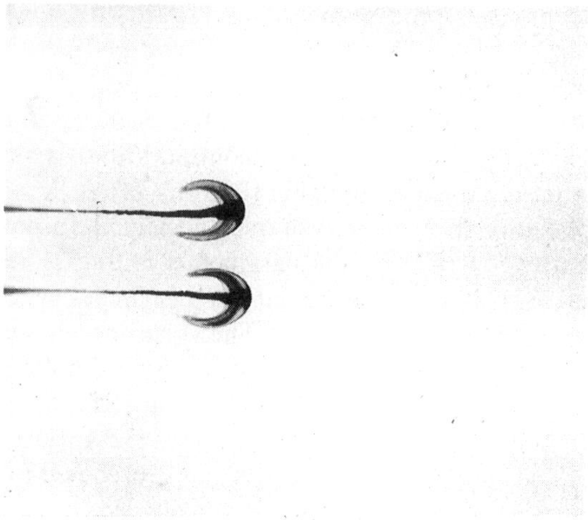

(a)

(c)

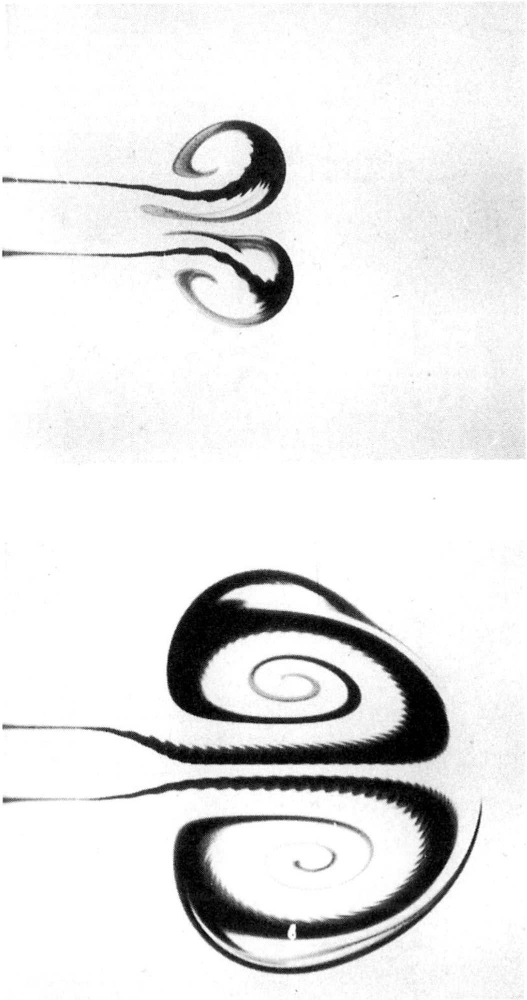

(b)

(d)

Figure 4.8 *Photographic sequence showing the interaction of two dipoles (*$\mathrm{Re} \approx 70$*) moving in parallel: (a)* $\mathrm{t} = 5.5$ *s; (b) 16 s; (c) 60 s; (d) 118 s. The distance between sources* $\varepsilon = 2$ *cm.*

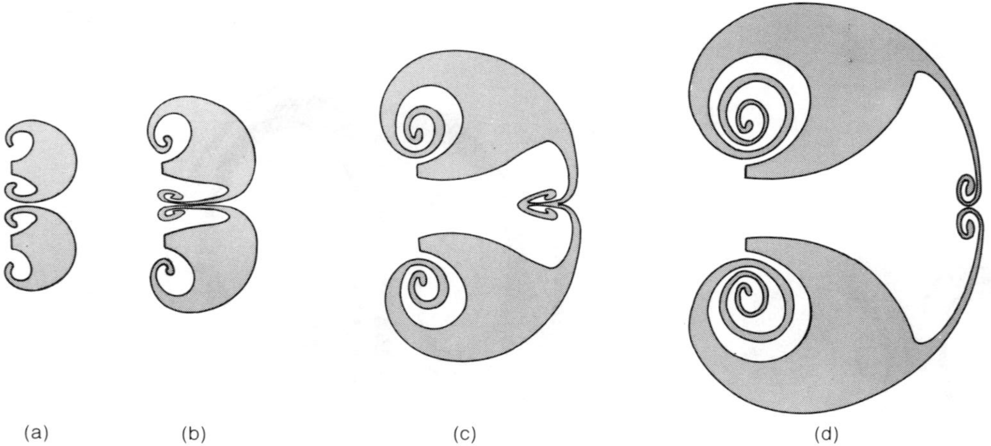

Figure 4.9 *Dyed water distributions for dipoles moving in parallel. For calculations a superposition of two solutions (3.5.15) was used:* $J_1 = J_2 = 0.04\,cm^3\,s^{-2}$, $\nu = 10^{-2}\,cm^2\,s^{-1}$, *the distance between sources* $\varepsilon = 0.5\,cm$, *and* $t = 3\,s$ *(a),* $5\,s$ *(b),* $10\,s$ *(c) and* $16\,s$ *(d).*

and two new dipoles form. The outer vortices form a new strong dipole, which moves forward. The inner vortices form a weak dipole, which moves in the opposite direction (Figures 4.10b–d). The initial phase is clear: two dipoles develop independently (Figure 4.10a). To explain the subsequent evolution of the flow, we must take into account that the dipole intensity (momentum) is a vector. Introducing the components of the momentum vector, we find that the flow shown in Figure 4.10 is governed by two parameters: $I = 2I \cos \frac{1}{2}\alpha$ and $M = \varepsilon I \sin \frac{1}{2}\alpha$, where α is the angle between the directions of the sources. Thus the flow represents the superposition of dipolar and quadrupolar flows. It is clear from the geometry that the induced velocities are in the same sense to the right of the collision point, and in opposite senses on the left-hand side. The outer vortices of the primary dipoles develop in the accompanying background dipolar flow and form a strong dipole (Figure 4.10b). The inner vortices develop against the background dipolar flow, forming a weak dipole (Figure 4.10b) which eventually decays (Figures 4.10c, d). Although it might appear surprising, the superposition of the linear solutions (3.5.15) correctly reproduces (Figure 4.11) the details of the flow evolution, including the weak dipole (Fig. 4.11d).

4.6 Non-axial collision of two dipoles

Suppose that two dipoles of equal momentum I move towards one another (Figure 4.12a), the distance between their axes being $\varepsilon \sin \alpha$ (the geometry is shown in Figure 3.11). As a result of interaction (Figures 4.12b, c), the inner vortices of the dipoles merge, forming a vortex core. The outer vortices form a periphery with vorticity of sign opposite to that of the core. The resulting S-shaped structure is clearly seen in Figure 4.12(d). Two parameters govern the flow dynamics: the quadrupole intensity $M = \varepsilon I$ of the flow and the total angular momentum $N = M \sin \alpha$ of the flow. Thus an unsteady rotating quadrupole (section 3.6.2) forms. Again using the linear solutions (3.5.15), one can calculate the dyed water distributions (Figure 4.13) for this type of interaction, and can observe the formation of a rotating quadrupole.

The S-shaped structure appears in experiments only for some ranges of the external governing parameters. In other cases it breaks

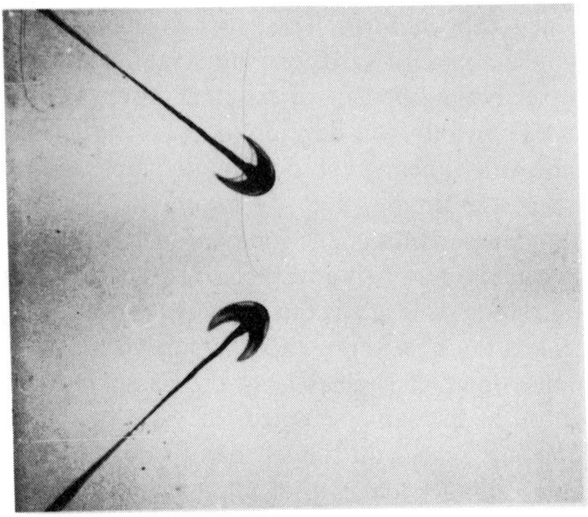

(a)

(c)

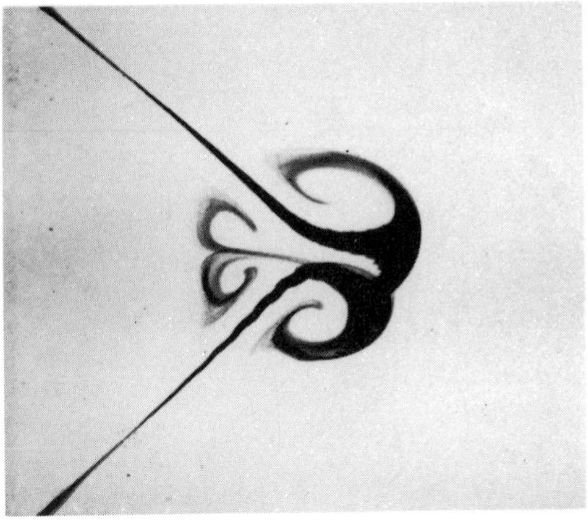

(b)

(d)

Figure 4.10 *Photographic sequence showing the oblique collision of two dipoles ($Re \approx 60$): (a) $t = 4.5$ s; (b) 21 s; (c) 32 s; (d) 56 s. The distance between sources $\varepsilon = 14$ cm.*

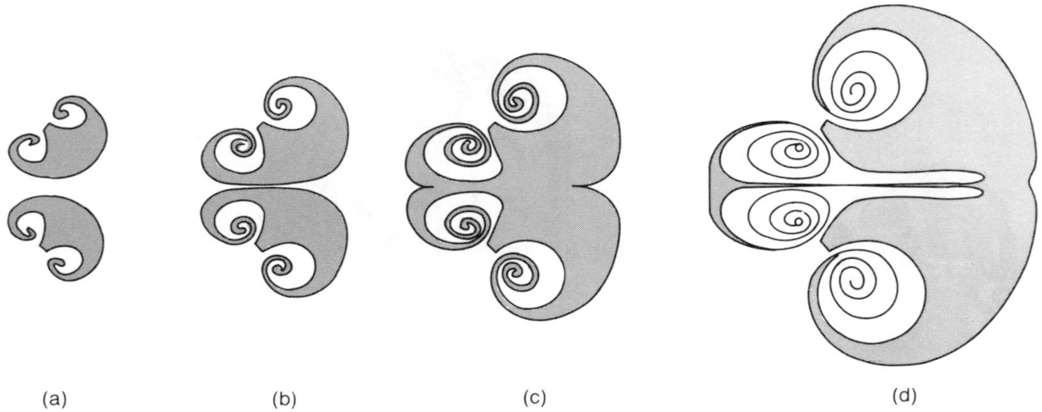

Figure 4.11 *Dyed water distributions for two dipoles colliding at an angle* $\alpha = \frac{1}{2}\pi$. *For calculations a superposition of two solutions (3.5.15) was used:* $J_1 = J_2 = 0.06\,cm^3\,s^{-2}$, $\nu = 10^{-2}\,cm^3\,s^{-1}$, *and* $t = 3\,s$ *(a),* $5\,s$ *(b),* $8\,s$ *(c) and* $15\,s$ *(d).*

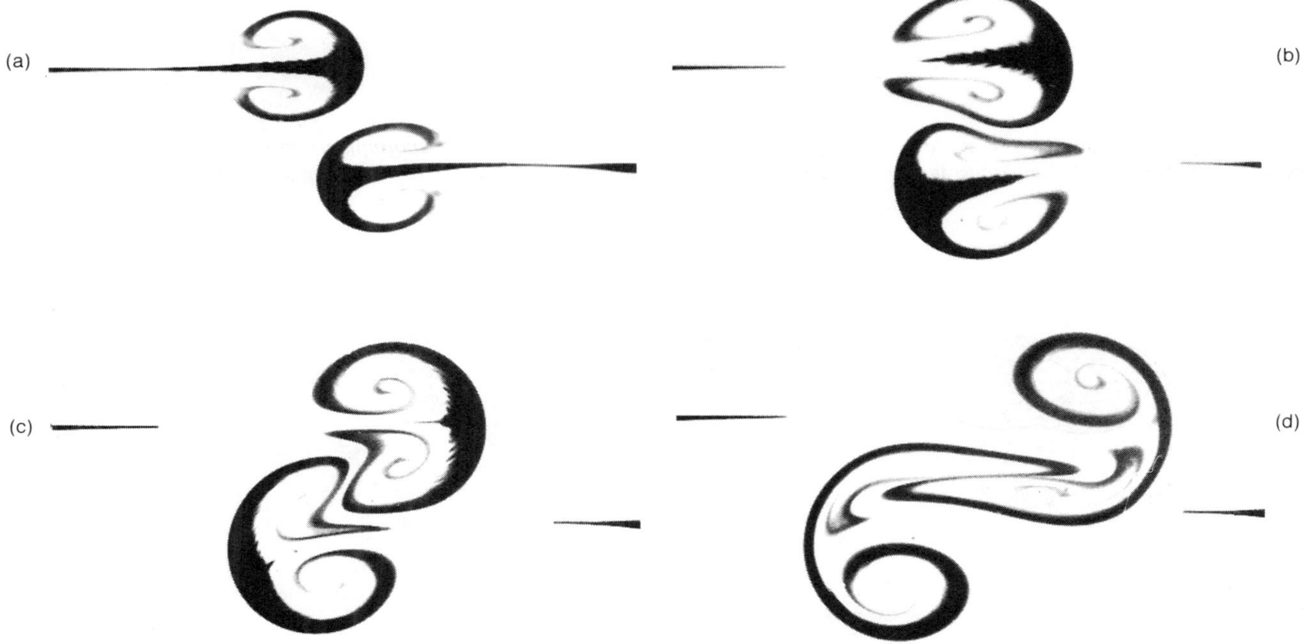

Figure 4.12 *Photographic sequence showing the non-axial interaction of two dipoles. The distance between sources* $\varepsilon = 7\,cm$ *and* $\alpha = 15°$ *(Fig. 3.11).*

(a) (b) (c) (d)

Figure 4.13 *Dyed water distributions for the non-axial interaction of two dipoles and the subsequent formation of an unsteady rotating quadrupole. For calculations a superposition of two solutions (3.5.15) was used:* $J = 0.04\,cm^3\,s^{-2}$, $\nu = 10^{-2}\,cm^2\,s^{-1}$, *and* $t = 1\,s$ *(a),* $3\,s$ *(b),* $6\,s$ *(c) and* $10\,s$ *(d).*

up into two dipoles. The transition criterion has not yet been determined.

4.7 Symmetric collision of a dipole with a wall

Suppose that a source of momentum of intensity I_0 is at a distance ε from a wall and is directed normally to the latter. Initially, a turbulent cloud impinges on the wall (Figure 4.14a), and eventually splits into two parts, forming two dipoles (Figure 4.14b). The newly formed dipoles, consisting of intense primary vortices and weaker satellites, rebound from the wall and move in circular paths (Figure 4.14c). The weaker satellites eventually decay. As a result, two large vortices of increasing radius move gradually apart along the wall.

If there were a free-slip condition at the wall, one would expect the situation to be similar to the symmetric collision of two dipoles of equal intensity (section 4.1). In that case the wall would act only as a mirror. But in a real fluid vorticity is created at the wall. Consequently, the flows for no-slip and free-slip conditions are different. When the source begins to act, the induced potential dipolar flow causes the emergence of frictional forces in the boundary layer at the wall. These forces generate two vortex patches of opposite sign (Figure 4.15). These patches can be considered as a newly formed dipole directed away from the wall.

To explain the subsequent evolution of the flow, consider the interaction of primary and newly formed dipoles. This stage of the process can be interpreted in the same way as the collision of two dipoles of different intensities (section 4.2). Thus, in addition to the main governing parameter – the momentum I_0 of the primary dipolar patch – we can introduce the momentum I_* of the dipolar vorticity distribution generated at the wall. We do not know the exact value of I_*, which is a function of the initial governing parameters and time, but clearly it cannot exceed I_0. Thus two parameters govern the flow dynamics: $I = I_0 - I_*$ and $M = I_*\varepsilon$. Owing to the presence of the virtual quadrupole, two dipoles, consisting of vortices from the split primary dipole and vortices detached from the wall, are formed. The newly formed dipoles then move in looping curves in the net background dipolar flow of intensity I (Figures 4.14b, c). The weaker satellites, generated at the wall, decay while orbiting around the primary vortices. As a result, a single final

(a)

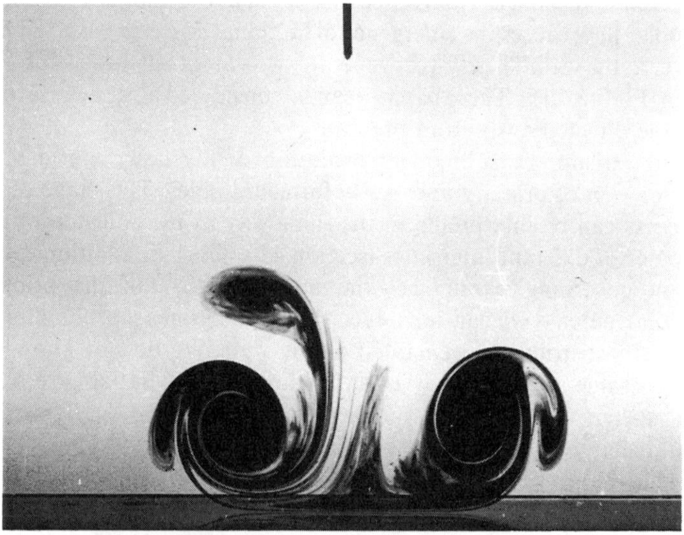

(c)

(b)

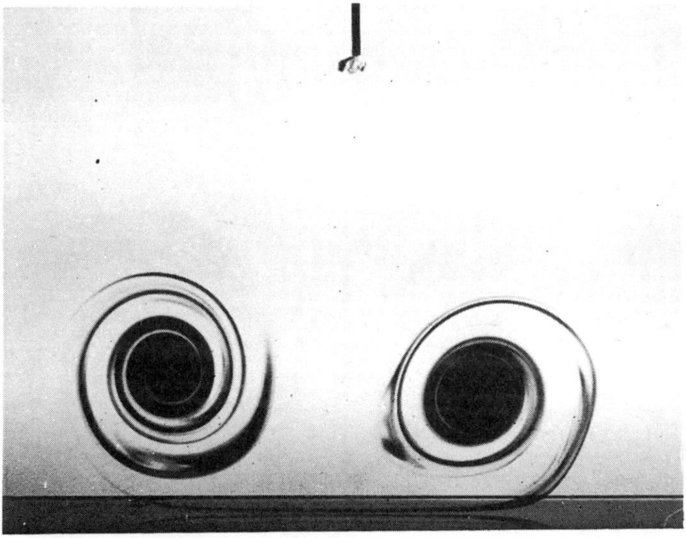

(d)

Figure 4.14 *Photographic sequence showing the collision of an impulsive strong jet ($Re \approx 700$) with a wall: (a) $t = 14$ s; (b) 51 s; (c) 75 s; (d) 136 s. The distance between the source and the wall $\varepsilon = 14$ cm.*

dipole is formed (Figure 4.14d) as if the process had returned to the very beginning, with one essential difference: the momentum I of the final dipole is smaller than that (I_0) of the primary dipole. Thus, by this process, the wall reduces the momentum of the flow.

Now consider the effect of image vortices on the flow dynamics. We can interpret this stage of interaction using the same arguments as for the symmetric collision of two dipoles (section 4.1). As a result, two vortices (half-dipoles) move apart along the wall. A numerical two-dimensional simulation by Orlandi (1990) has shown that the number of intermediate bifurcations and the number of secondary vortices created at the wall increase when the initial Reynolds number of the flow (momentum of primary dipole) is increased. In the experiment the initial Reynolds number was not very large, and only one bifurcation occurred. The current Reynolds number of the impulsive vortex dipole decreases with time (section 5.2). Thus after a number of bifurcations the nonlinear effects become small. Hence the secondary vorticity generated at the wall does not become concentrated into localized patches, and the process of viscous diffusion prevails. Therefore, for moderate initial Reynolds numbers,

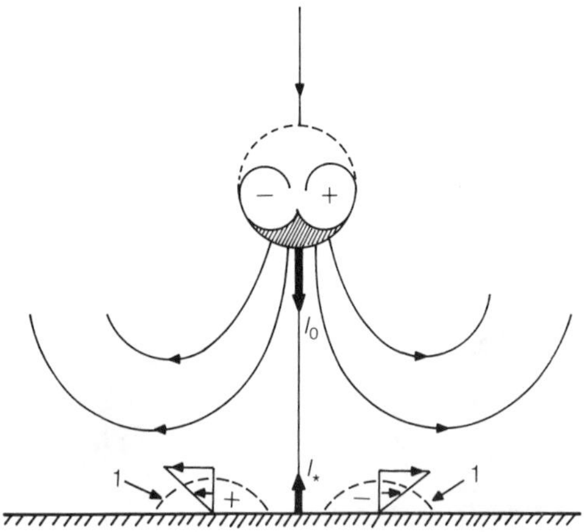

Figure 4.15 *Sketch of the initial stage of the collision of a vortex dipole with a wall. Secondary vortex patches (1) are created in the boundary layer at the wall by the flow induced by the dipole. The two heavy arrows show schematically the momenta* I_0 *and* I_* *of the primary and newly formed dipole vortex patches.*

the final result of the symmetric collision is the generation of two vortices of increasing size moving apart with decreasing velocity along the wall (Figure 4.14).

4.8 Parallel motion of a dipole along a wall

Consider the case when the source of momentum I_0 is at some small distance ε from the wall and acts parallel to the latter. The photographic sequence in Figure 4.16 shows the evolution of a dipolar vortex patch that moves along the wall. It can be seen that the dipole forms (Figure 4.16b) in a turbulent cloud and then emerges from the wall (Figure 4.16c). The eruption of the dipole from the wall resembles the bursting phenomenon in the near-wall region of a turbulent boundary layer (Falco, 1977; Robinson, 1991). The outer vortex stretches and eventually decays. A single vortex (half of a dipole) of increasing size, moving forward with decreasing velocity, is the result of the interaction (Figure 4.16f).

To explain the evolution of the flow shown in Figure 4.16, consider first a dipolar vortex patch of intensity I_0 moving along the wall. As a consequence, a thin boundary layer forms at the wall. The dipolar vortex cloud moving along the wall sweeps the vorticity from the boundary layer. This leads to an asymmetry of the vorticity distribution in the cloud. At the same time the interaction of the patch with its mirror image occurs. The subsequent evolution of the flow resembles the interaction of two dipoles moving in parallel and separated by a distance 2ε (section 4.4). The result of this interaction is a compact vortex (half of a dipole), which slowly moves along the wall with decreasing velocity. Thus the wall in the experiment of Figure 4.16 is more than just a symmetry plane: its viscous boundary layer causes the vortex to move away from the wall over a considerable distance (compare Figures 4.16f and 4.8d).

The question arises, where has the vorticity from the decaying vortex gone? A natural hypothesis is that all of this vorticity goes into the vortex tail and then effectively diffuses.

4.9 Collision of a dipole with a small body

To demonstrate the crucial role of the secondary vorticity created at a solid boundary because of the no-slip condition, consider the

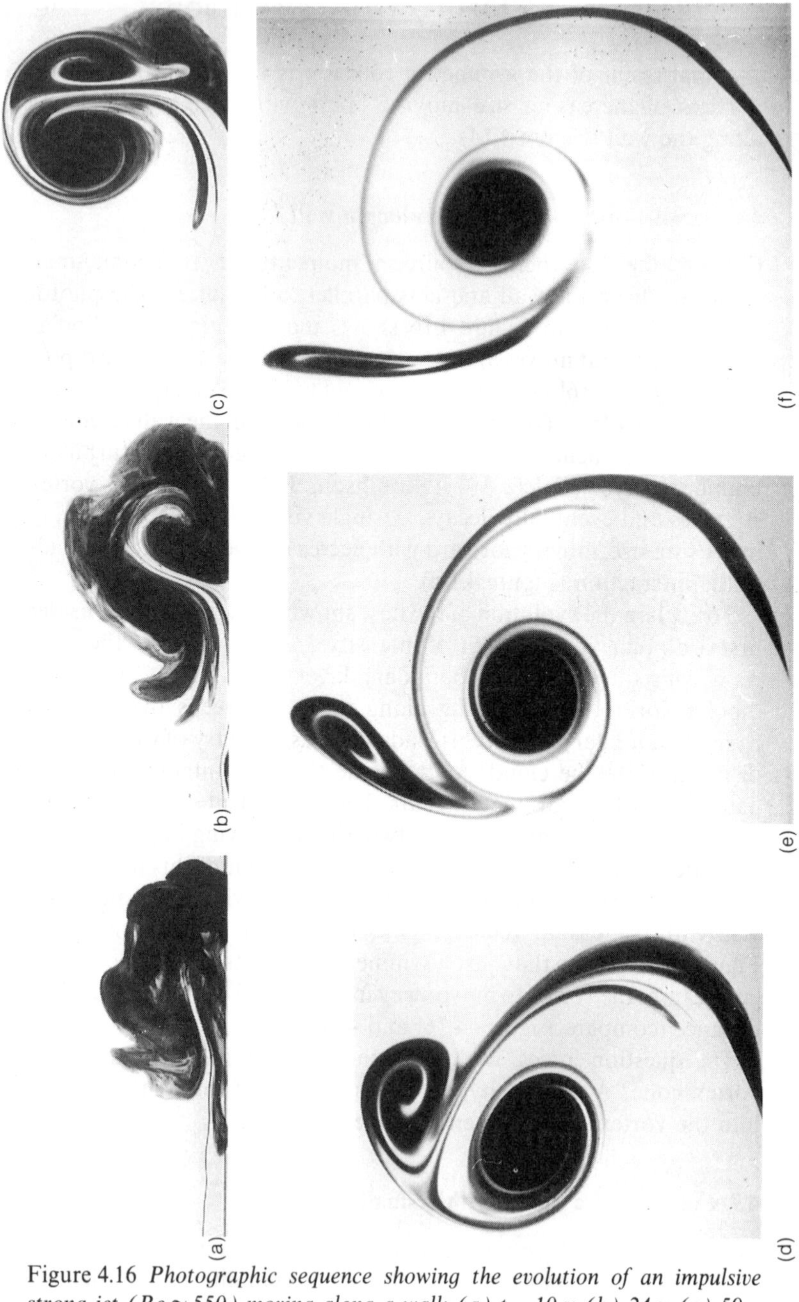

Figure 4.16 *Photographic sequence showing the evolution of an impulsive strong jet (Re ≈ 550) moving along a wall: (a) t = 10 s; (b) 24 s; (c) 59 s; (d) 162 s; (e) 320 s; (f) 530 s. The distance between source and wall ε = 1 cm.*

interaction of a dipole with a body whose horizontal dimension is much less than the size of the dipole (Figure 4.17). It turns out that the interaction dynamics closely resemble the interaction of two dipoles of different intensities (Figure 4.4), and can be interpreted in a similar manner. The similarity between these two cases is not only visual, but also has a physical basis. Because of the no-slip condition, the primary dipolar flow (I_1) acts on a body with some force; hence the body acts on the ambient fluid with the same force in the opposite direction. For a small body this force can be considered to be localized. The action of a localized force on a viscous fluid of necessity gives rise to the formation of a secondary dipolar flow (I_2) directed against the primary flow. Thus, qualitatively, we arrive at the case of the head-on collision of two dipoles of different momenta (section 4.2). In both cases (compare Figures 4.4 and 4.17) the final result of the interaction is a single dipole of reduced intensity.

In general, the intensity I_2 of the secondary dipole is a complicated function of time, position, the geometry of the body and the intensity I_1 of the primary dipole. However, in some cases, when the primary flow is well documented and the body has a simple geometry, the value of I_2 can be estimated. Consider the interaction of a continuously generated dipole (developing jet in a stratified fluid) with a small vertical cylinder (Figure 4.18). The distribution of longitudinal velocity u in this primary flow of intensity J_1 (just after the frontal dipole) is given by (2.3.29). A typical half-width r_0 of this distribution can be found from (2.3.29), and at $x = \varepsilon$ (where ε is the distance from the body to the origin) $r_0 \approx 8\varepsilon\nu/J_1^{1/2}$. For a small body with $a \ll r_0$ (where $2a$ is the diameter of the body) the flow running over the body is almost uniform. For a cylinder placed in a uniform flow of amplitude u the drag force (per unit length) for moderate values of the Reynolds number ($Re_a = 2au/\nu$) is given by (e.g. Batchelor, 1967)

$$\rho F \approx 10\rho u^2 a/Re_a^{1/2}. \tag{4.9.1}$$

Integrating this over the vertical coordinate z with u given by (2.3.29), we obtain an estimate of the total force ρJ_2 applied by the cylinder:

$$J_2 = \int_{-\infty}^{\infty} F\,\mathrm{d}z = \frac{45}{16}\left(\frac{a}{\varepsilon}\right)^{1/2} J_1.$$

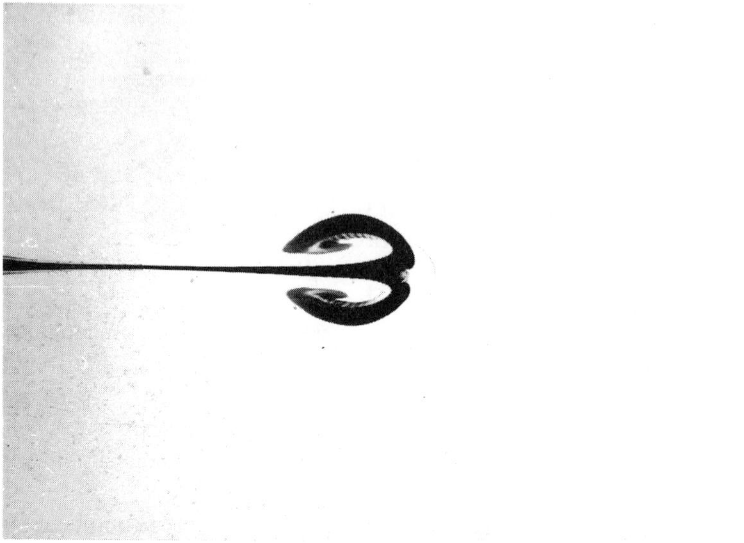

(a)

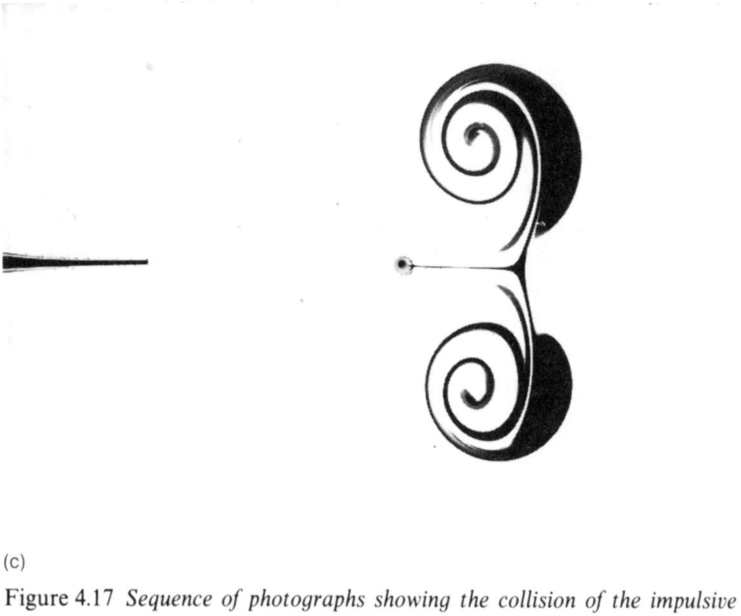

(c)

Figure 4.17 *Sequence of photographs showing the collision of the impulsive dipole ($\Delta t = 5$ s, $Re \approx 50$) with a small vertical cylinder: (a) $t = 2$ s; (b) 9 s;*

(Continued on next page)

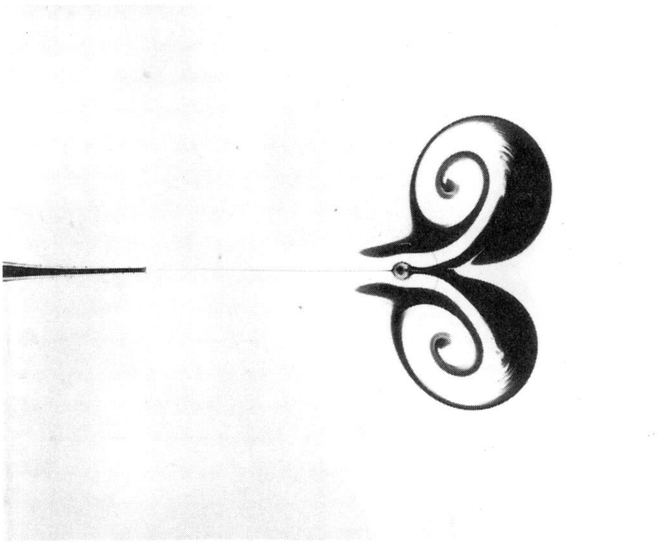

(b)

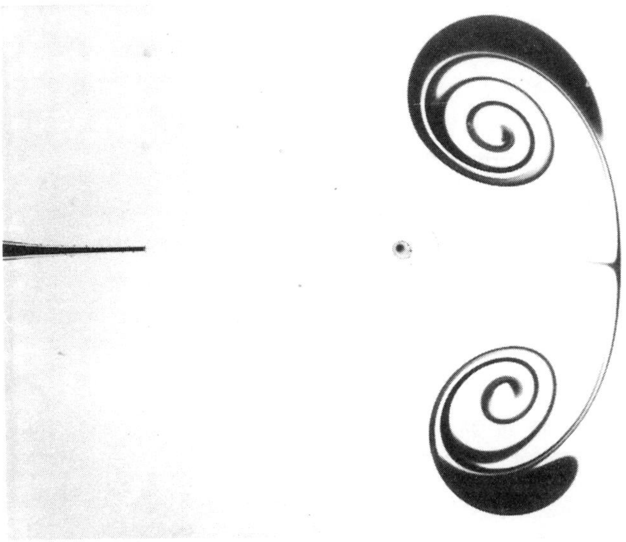

(d)

(c) 17 s; (d) 48 s. The diameter of the cylinder $2a = 0.125$ cm, the distance from the nozzle $\varepsilon = 4.4$ cm. Note that the cylinder is much smaller than its support visible in the photographs.

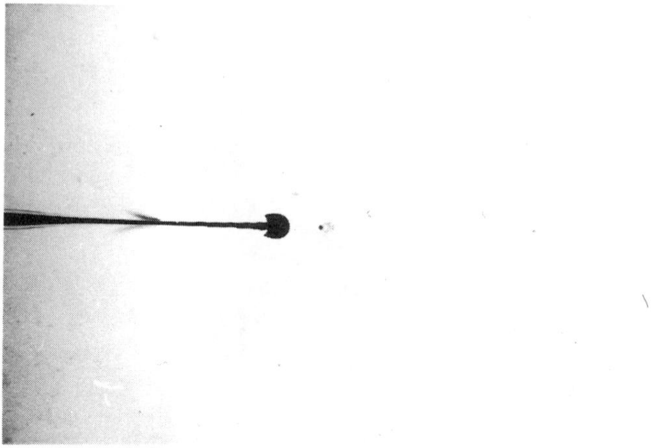

(a)

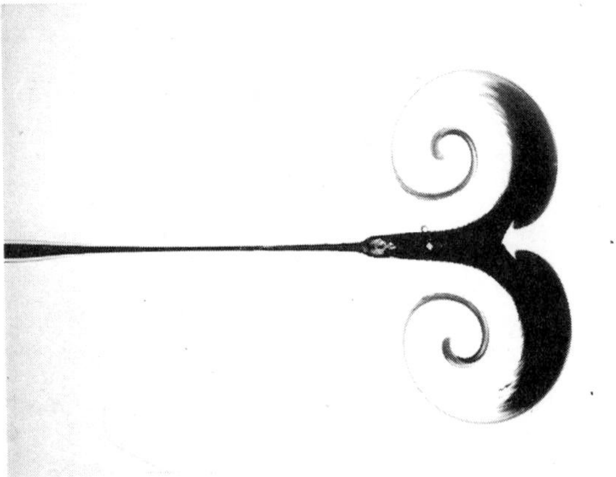

(c)

Figure 4.18 *Sequence of photographs showing the central collision of a continuously generated dipole in a linearly stratified fluid with a vertical*

(*Continued on next page*)

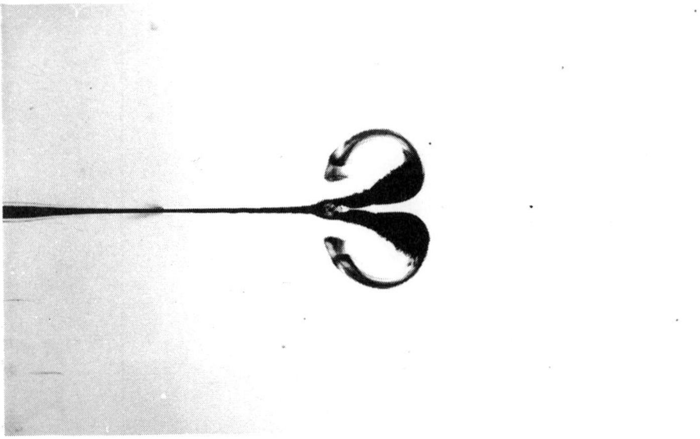

(b)

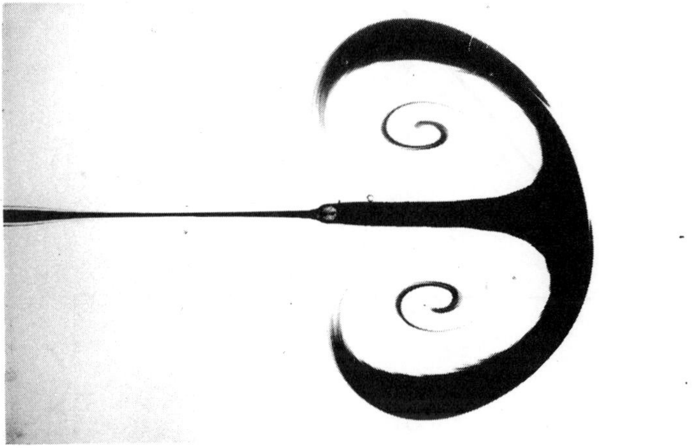

(d)

cylinder: (a) $t = 1$ *s; (b) 6 s; (c) 10 s; (d) 31 s.* $J_1/\nu^2 = Re^2 = 4300$, $2a = 0.125$ *cm,* $\varepsilon = 3.3$ *cm and* $\mathcal{N} = 2\,s^{-1}$.

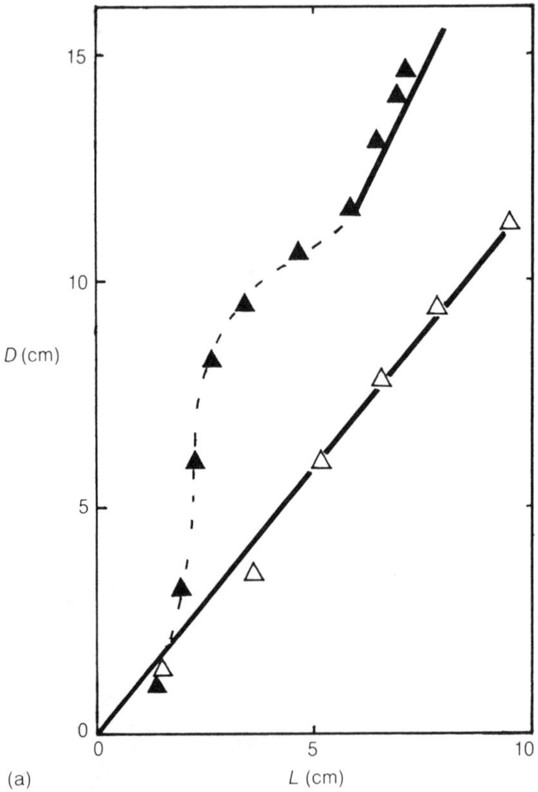

(a)

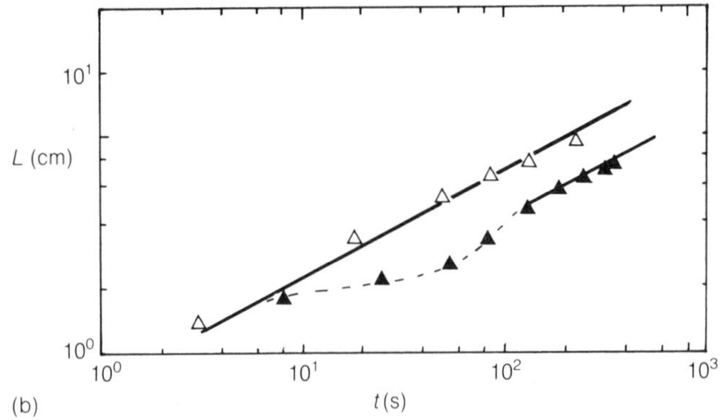

(b)

Figure 4.19 *Comparision of measured (symbols) and estimated (solid lines) values of D for different L (a) and of L for different t (b) for two experiments with the same forcing* $J_1/\nu^2 = 1385$ *and* $\mathcal{N} = 2\,s^{-1}$. *Δ, the experiment without a cylinder;* ▲, *the experiment with a cylinder* ($2a = 0.125$ *cm,* $\varepsilon = 2$ *cm*). *The dashed lines represent the intermediate stage of interaction and are shown only for convenience.*

Thus after the interaction the newly formed dipolar flow has the reduced intensity

$$J = J_1 - J_2 = J_1\left[1 - \frac{45}{16}\left(\frac{a}{\varepsilon}\right)^{1/2}\right]. \tag{4.9.2}$$

The model that will be considered in section 5.1 allows one to estimate the horizontal cross-width $2R = D$ of the frontal dipole in the developing jet, as well as the distance L from the origin to the centre of the dipole: these flow properties increase with time as $D \propto L \propto t^{1/2}$, with the coefficients of proportionality being universal functions of only the non-dimensional force intensity J/ν^2 (Figure 5.6). Thus, for given J_1, one can calculate J by means of (4.9.2) and then estimate D and L. The values of D and L measured in two experiments (without and with a cylinder) and the estimates obtained by the above-mentioned method are shown in Figure 4.19. It can be seen that in the intermediate stage of the interaction a bifurcation occurs, and a smooth peak and gap appear in the graphs. Finally, after the interaction, the properties of the newly formed dipole exhibit a different asymptotic behaviour from that of the free flow. This

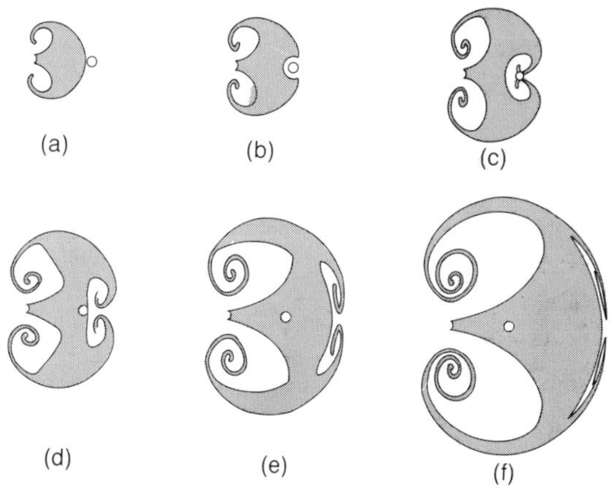

Figure 4.20 *Dyed water distributions calculated for the superposition of two solutions (3.5.15): $J_1 = 0.06\ cm^3\ s^{-2}$, $J_2 = \frac{1}{2}J_1$, $\nu = 10^{-2}\ cm^2\ s^{-1}$, $\varepsilon = 0.5$ cm, and t = 1.9 s (a), 2.7 s (b), 4.2 s (c), 5.3 s (d), 7.8 s (e) and 11.8 s (f).*

behaviour corresponds to a dipole of smaller intensity J than the intensity J_1 of the primary dipole: after the interaction the newly formed dipole moves more slowly and its width increases more rapidly with distance, in accordance with the estimates shown by solid lines in Figure 4.19.

Again using the superposition of two solutions (3.5.15), one can model the interaction process mathematically. Two point forces act on one another with intensities J_1 and J_2, the distance between them being ε. The first force reproduces the primary dipolar flow, while the second models the action of a small body on the primary flow. In Figure 4.20 the right-hand source models a body and is shown by a small circle. It starts when the dyed front of the primary flow reaches the point at which it is placed, and stops after some period of time. As can be seen, the model reproduces qualitatively correctly the process of interaction: splitting of the primary flow into two dipoles (Figures 4.20c, d) and the formation of a new dipole of reduced intensity at the end of this stage (Figures 4.20e, f).

5

Empirical models of vortex structures in a stratified fluid

The analytical solutions considered in Chapter 3 describe the creation of vortex multipoles induced by an appropriate localized forcing. These solutions also give useful information about the motion of passive tracers in such flows. However, the planar Stokes solutions as well as the weakly nonlinear solutions fail to describe quantitatively the real flow in a stratified fluid, and there are very few possibilities of significantly improving the solutions obtained. The exact self-similar solutions for vortex monopoles (3.2.3) and (3.3.7), which satisfy the Navier–Stokes equation, are rare exceptions. For multipoles of higher order it is possible to calculate the second, nonlinear, term in the Reynolds number expansion, but only for 'weak' multipoles. To make some progress in a quantitative description of real unsteady nonlinear three-dimensional flow in a stratified fluid, we must reconsider and simplify the problem. Theoretical analysis and experimental results demonstrate that a vortex structure generated by localized forcing can be considered as a compact vortex. This means that one can distinguish a finite region in the fluid within which the motion is essentially vortical and outside of which the motion is nearly irrotational. Thus the vortex structure can be related to some compact fluid volume. Generally, this volume moves in the fluid and its size increases. If we do not intend to consider the initial evolution of the flow and the details of motion in the vortex structure, we can attempt to describe the dynamics of the associated fluid volume, disregarding the particular nature of the flow. This description is usually based on integral relations expressing the balance of some essential physical quantities (e.g. mass, momentum or buoyancy) for the fluid volume under study. This approach was used by Maxwell for a vortex ring, and since then it has been employed successfully by many authors, in particular by Morton (1959) and Turner (1966, 1969) to describe the dynamics of plumes and thermals – buoyant volumes formed when a fluid is heated locally.

Success in using this method depends on an appropriate parameterization of external fluxes of mass and other conserved quantities into the moving fluid volume. On the basis of some physical hypothesis, the mass and momentum fluxes can be connected with the main external parameters governing the flow dynamics. As a result, a closed system of integral relations for the main global properties of the considered volume can be obtained. The correctness of the hypothesis used in the model can be justified only by comparison with experimental results. That is why this method is closely connected with experiment and why such models are called empirical.

In the following sections this approach is used, and empirical models for the vortex dipole and quadrupole are discussed. The results of calculations and estimates of some global properties of the flow are compared with data obtained in laboratory experiments. (For more details see Voropayev and Filippov (1985), Voropayev, Afanasyev and Filippov (1991) and Voropayev and Afanasyev (1993b)).

5.1 A developing horizontal jet in a stratified fluid

To produce experimentally a submerged jet in a linear stratified fluid, a thin round nozzle from which a fluid is injected is used as a localized source of momentum. When the source is turned on, an almost-spherical vortex appears, similar to that in a homogeneous fluid (Figure 5.1). Initially, the vortex develops as in a homogeneous fluid because the inertial force exceeds the gravitational force. With increasing distance, the latter begins to prevail, and the initially spherical vortex becomes flat and expands horizontally, forming a frontal vortex dipole (Figure 5.2). After that the shape of the vortex front does not change significantly, and the flow develops self-similarly up to a large distance from the origin.

A developing flow with a continuously acting momentum source J can be represented as consisting of two parts: a steady jet region and a frontal region moving forward with velocity $\bar{U}$. Consider a physical model based on the scheme shown in Figure 5.3. The frontal

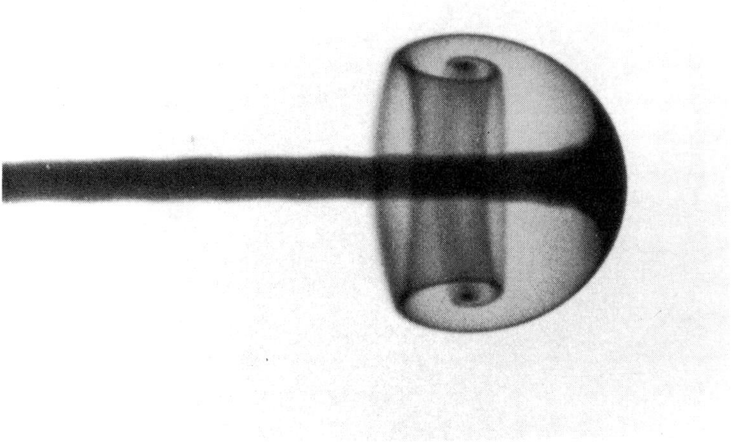

Figure 5.1 *Dyed starting jet in a homogeneous fluid at $Re = 55$.*

region as a whole moves with a velocity $\bar{U}$ that is less than the local velocity U_0 of the fluid in the jet behind the front. So the fluid from the jet always flows into the frontal region. This fluid then spreads horizontally, forming a vortex dipole. Observations demonstrate (Figure 5.2 and 5.4) that the frontal region looks like a flat disc with horizontal cross-width $D = 2R$ and mean vertical thickness H. The front has a streamlined shape and moves with velocity

$$\bar{U} \equiv \frac{\mathrm{d}L}{\mathrm{d}t}, \tag{5.1.1}$$

where L is the distance of the centre of the frontal region from the origin. Thus at all times the flow can be characterized by three global parameters: L, D and H. In a self-similar flow all of these length scales must be proportional to each other. Let us assume that

$$D = \alpha L, \tag{5.1.2}$$

$$H = 2\gamma L, \tag{5.1.3}$$

where α and γ are non-dimensional coefficients that do not depend on x or t.

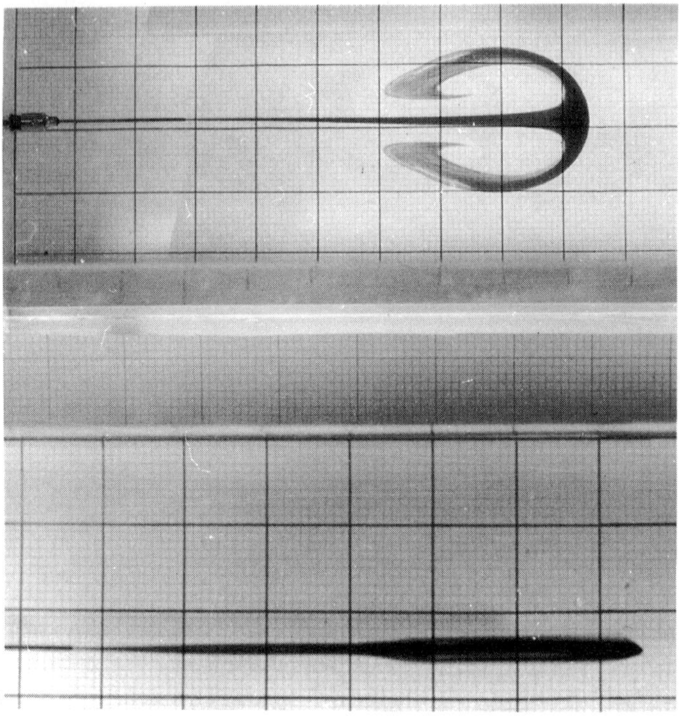

(a)

(b)

Figure 5.2 *Dyed, developing horizontal jet in a stratified fluid: (a) top view; (b) side view.* $Re = 85$, $\mathcal{N} = 1.2\,\mathrm{s}^{-1}$; *the grid is in inches.*

The results of measurement demonstrate that the distance L increases with time as $t^{1/2}$. Hence the velocity $\bar{U}$ varies as

$$\bar{U} = \frac{\mathrm{d}L}{\mathrm{d}t} \propto t^{-1/2} \propto x^{-1},$$

and thus varies with distance in the same way as the velocity U_0 at the axis of a steady jet in a stratified fluid, (2.3.28). Thus we can write

$$\bar{U} = \beta U_0 \quad (x = L), \tag{5.1.4}$$

where β is also a non-dimensional coefficient that does not depend on x. We want to calculate the values of the coefficients α, β and γ. The agreement between the theoretical and experimental values of these coefficients will be a criterion for the correctness of the model.

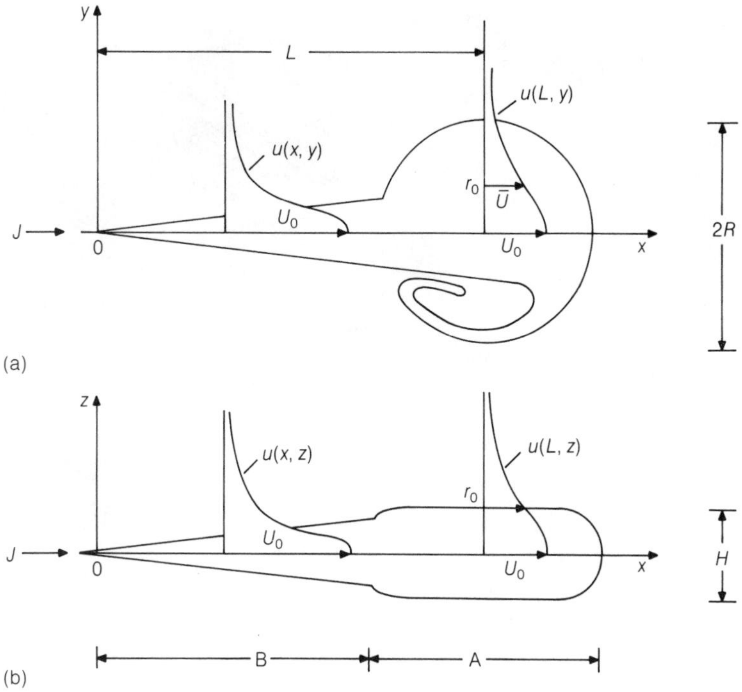

Figure 5.3 *Sketch of a developing jet in a stratified fluid and coordinate system: (a) top view; (b) side view; A = front region; B = jet.*

Let us consider the frontal region as a fluid volume V moving forward with velocity $\bar{U}$. Assume that the jet that flows into the rear of this region has a velocity distribution $u(x, y, z)$ given by (2.3.29), and transports some momentum and mass into the frontal region. The mass and momentum balances for the fluid volume V are

$$\frac{\mathrm{d}V}{\mathrm{d}t} = 2\pi \int_0^{r_0} (u - \bar{U})\, r\, \mathrm{d}r + q, \tag{5.1.5}$$

$$(1 + k)\frac{\mathrm{d}(\bar{U}V)}{\mathrm{d}t} = 2\pi \int_0^{r_0} u(u - \bar{U}) r\, \mathrm{d}r - F, \tag{5.1.6}$$

where $k = 1$ is the virtual mass coefficient for a liquid disc. The drag force ρF and the additional volume influx q will be defined below.

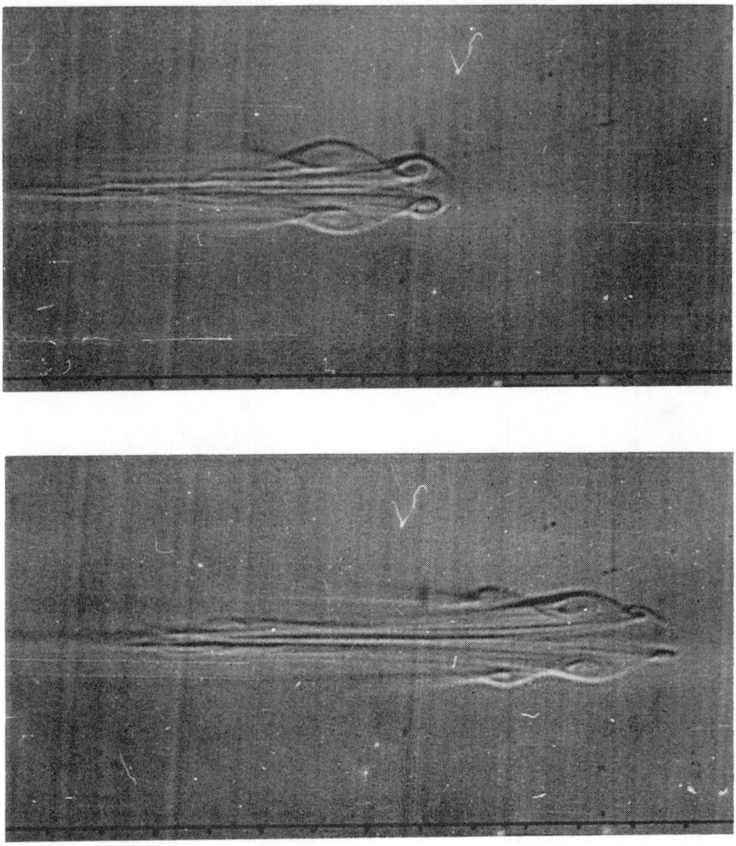

(a)

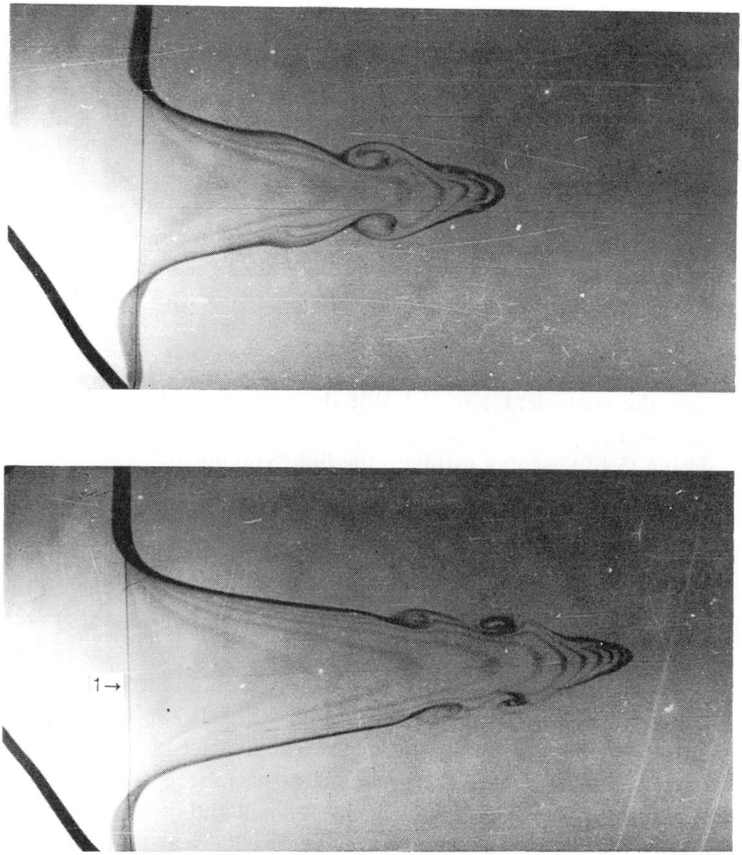

(b)

Figure 5.4 *Frontal region and part of the jet behind it (side view): (a) shadowgraph picture; (b) thymol blue visualization; 1 = platinum wire ($x = 17$ cm, $y = 0$) $\mathcal{N} = 1.5\ \mathrm{s}^{-1}$, Re = 104; the scale is in centimetres.*

The condition determining the distance r_0 from the jet axis at which the velocity in the jet u is equal to the frontal region velocity $\bar{U}$ is

$$u(r_0, L) = \bar{U}(L). \tag{5.1.7}$$

Using (2.3.29), (5.1.1) and (5.1.4), we obtain from (5.1.7)

$$r_0 = (1 - \beta^{1/2})^{1/2} L B^{-1/2} \beta^{-1/4} J^{-1/2} \nu. \tag{5.1.8}$$

Fluid particles with $r > r_0$ move more slowly than the frontal region, and never enter it. Thus an estimate of the mean value of the frontal region thickness H is

$$H = 2r_0. \tag{5.1.9}$$

Hence the coefficient γ in (5.1.3) is determined, and we shall now seek α and β.

Using (5.1.9), we can estimate the fluid volume as

$$V = 2\pi R^2 r_0.$$

Now consider a flat fluid disc of radius R and height H, intruding horizontally into the linearly stratified resting fluid. In general, the drag force F in (5.1.6) is the sum of wave (F_w) and viscous (F_v) drag forces.

Typical 'instantaneous' salinity profiles across the frontal and jet regions of the flow are shown in Figure 5.5(a). It can be seen that the fluid particles in the frontal region are displaced from the plane $z = 0$. In contrast, the fluid particles in the jet region are sucked towards this plane. This displacement is small and decreases rapidly with distance from the origin. Thus the density distributions in the flow when x is not small are practically the same as in the ambient fluid. The time record shown in Figure 5.5(b) demonstrates that there are no significant internal waves in the flow. Small changes in the salinity on this record are due to small displacements of fluid particles in the jet and frontal regions from their equilibrium levels. Hence the amplitudes of the internal waves are very small and the wave drag F_w is negligible.

The viscous drag on a disc is determined mainly by the friction on its horizontal surfaces, and we can use the estimate

$$F \approx F_v = c\nu R\bar{U}, \tag{5.1.10}$$

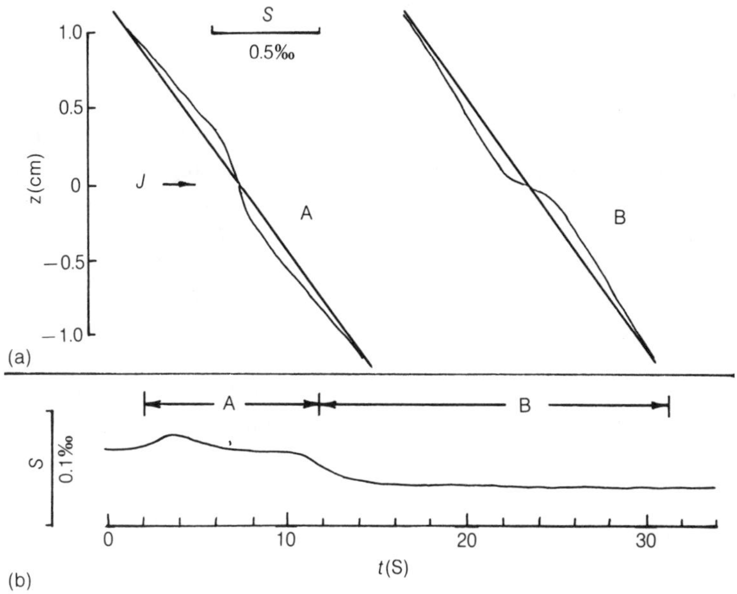

Figure 5.5 *(a) Vertical profiles of salinity in a developing jet; x = 5 cm and y = 0. (b) Time variation of salinity at a point. A = frontal region; B = jet; x = 5 cm, y = 0 and z = 1 cm. Re = 170. The straight lines in (a) show the undisturbed profile.*

where $c = \frac{32}{3}$ (Lamb, 1932). The leading edge of the starting jet, intruding into the resting fluid, splits up into two symmetric 'half-jets', which then wrap into spirals, forming the disc-like shape of the front region. On the outer boundary of these half-jets the initially quiescent ambient fluid forms a viscous boundary layer. Some fluid from this boundary layer is entrained into the frontal region, giving the additional influx q in the mass conservation equation (5.1.5). The total flux q_0 in a steady round jet increases linearly with distance x from the origin: $q_0 = 8\pi\nu x$ (Schlichting, 1979). One can interpret the steady jet as a free boundary layer and estimate the term q using this formula. In the present case the distance x must be measured from the splitting point, and it is equal to the arclength of the disc-like frontal region: $x \approx \pi R$. For two 'half-jets' we have the estimate

$$q \approx \lambda 8\pi^2 \nu R.$$

In a self-similar flow λ must be a constant. Its value is equal to the portion of fluid that enters from the boundary layer (on the outer boundary of the frontal region) into the frontal region, and $\lambda < 1$. In our model the value of λ is not known, but must be found from a comparison of calculated and measured values of α and β.

Thus the system of governing equations is closed. Without taking account of the exact initial conditions (nozzle size and mass flux ρq_* at the nozzle exit), one can seek solutions in the self-similar form (5.1.2) and (5.1.4). Recall that we have already determined the coefficient γ in (5.1.3). Integrating (5.1.5) and (5.1.6) for solutions in the form of (5.1.2) and (5.1.4), we have two algebraic equations for α and β:

$$\frac{3^{5/2}}{8\pi^{1/2}} \alpha^2 \beta^{3/4}(1-\beta^{1/2}) J^{1/2} \nu^{-1} = 2(1-\beta^{1/2})^2 + \lambda\alpha\pi, \tag{5.1.11}$$

$$\frac{3^{3/2}2}{\pi^{1/2}}(1+k)\alpha^2\beta^{7/4}(1-\beta^{1/2})^{1/2} J^{1/2}\nu^{-1} = \frac{2^4}{3}(1-3\beta+2\beta^{3/2}) - c\alpha\beta. \tag{5.1.12}$$

In (5.1.11) and (5.1.12) λ is a free parameter. Its value ($\lambda = 0.3$) was chosen by fitting of the experimental and calculated values of α and β. It turns out that the exact value of the drag force coefficient c in (5.1.10) and the virtual mass coefficient k do not enter very sensitively in the calculations. Thus the influence of the drag force on the flow dynamics is negligible. The momentum flux transported by the jet into the frontal region goes mainly into the acceleration of the initially quiescent fluid entrained into this region. The calculated and experimental values of α and β are shown in Figure 5.6 as functions of $J/\nu^2 = Re^2$. Knowing β, we can verify the relation (5.1.9). The results of the measurements are shown in Figure 5.7 in non-dimensional form: the values of r_0 for different L and J/ν^2 were calculated using (5.1.8) and the graph in Figure 5.6.

Introducing the Reynolds number of the frontal region

$$\overline{Re} = \frac{2R\bar{U}}{\nu} = \frac{\alpha\beta AJ}{\nu^2}, \qquad A = 3/8\pi, \tag{5.1.13}$$

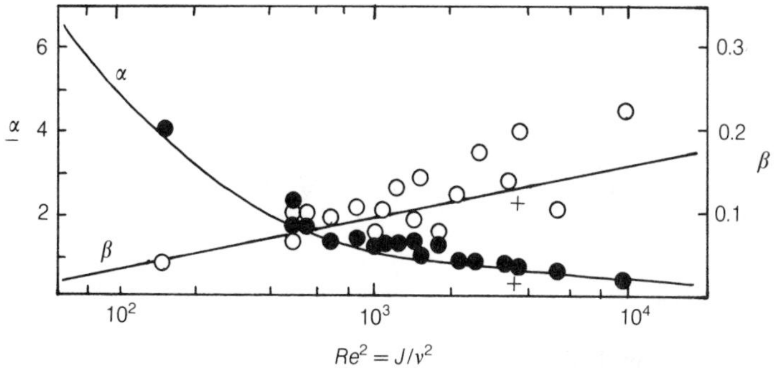

Figure 5.6 *Measured (symbols,* $\mathcal{N} = 0.5$–$2.5\ s^{-1}$*) and calculated (from (5.1.11) and (5.1.12),* $\lambda = 0.3$*, solid lines) values of* α *(●) and* β *(○) for different* $J/\nu^2 = Re^2$*. Crosses show the results of a three-dimensional numerical simulation by Voropayev and Neelov (1991).*

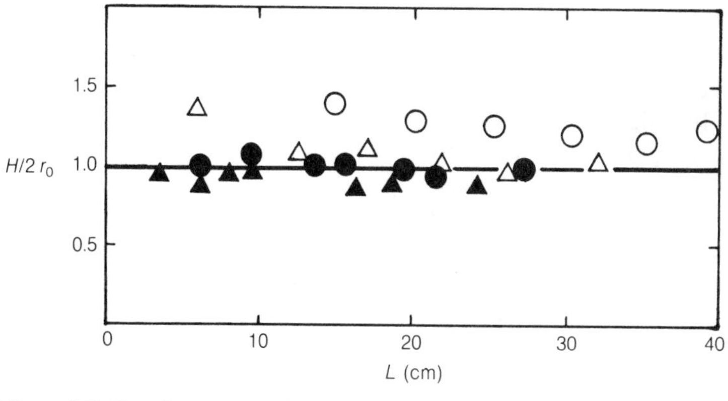

Figure 5.7 *Non-dimensional thickness* $H/2r_0$ *of the frontal region for different distances L from the origin and* $\mathcal{N} = 1.4\ s^{-1}$*: ▲,* $Re = 56$*; ●, 73; Δ, 107; ○, 146.*

we can write some useful estimates

$$L = \left(\frac{2\overline{Re}}{\alpha}\right)^{1/2} (\nu t)^{1/2}, \tag{5.1.14}$$

$$R = \left(\tfrac{1}{2}\alpha\,\overline{Re}\right)^{1/2} (\nu t)^{1/2}, \tag{5.1.15}$$

$$\bar{U} = \left(\frac{Re}{2\alpha}\right)^{1/2}\left(\frac{\nu}{t}\right)^{1/2}, \qquad (5.1.16)$$

$$H = 8\beta^{1/4}(1-\beta^{1/2})(\nu t)^{1/2}. \qquad (5.1.17)$$

From (5.1.14) and (5.1.16) we have a simple estimate of the 'age' of the flow:

$$t = L/2\bar{U}. \qquad (5.1.18)$$

Thus, on the basis of the idea that in the intermediate–asymptotic stage the flow develops in a self-similar fashion, we have obtained estimates of the major flow properties.

5.2 An impulsive vortex dipole

Consider a localized source of momentum, acting, in contrast to the previous case, for a short period of time Δt, injecting impulsively a strong horizontal jet into a density-stratified fluid. The jet transports a horizontal momentum $\rho I = 4\rho q_*^2 \Delta t/\pi d^2$ into the fluid. Here ρq_* is the mean mass flux at the nozzle exit, whose diameter is d (section 1.2.3).

In a density-stratified fluid the vertical motion is suppressed by gravity, and the flow rapidly becomes horizontal. The frontal region of the flow moves with a velocity $\bar{U}$ that is less than the local velocity U_0 of the fluid in the flow behind the front. The vorticity generated by the momentum source is transported by the jet-like flow into the frontal region, and is concentrated there, forming two flat vortices of opposite signs – a vortex dipole (Figure 1.3). The formation process finishes in a time of approximately $t_0 \approx \Delta t/(1-\beta)$, when the main part of the moving fluid enters the frontal region. Here

$$\beta(Re) = \frac{\bar{U}}{U_0}, \qquad 0 < \beta < 1, \qquad Re = \frac{4q_*}{\pi d\nu},$$

where Re is the initial Reynolds number of the jet-like flow.

The vortex dipole thus formed moves forward, and its horizontal dimension increases through the entrainment of ambient fluid. Vertical profiles of salinity in the flow before and after the vortex dipole has been formed are shown in Figure 5.8. Note that the density distribution in the flow before and after the vortex dipole has been

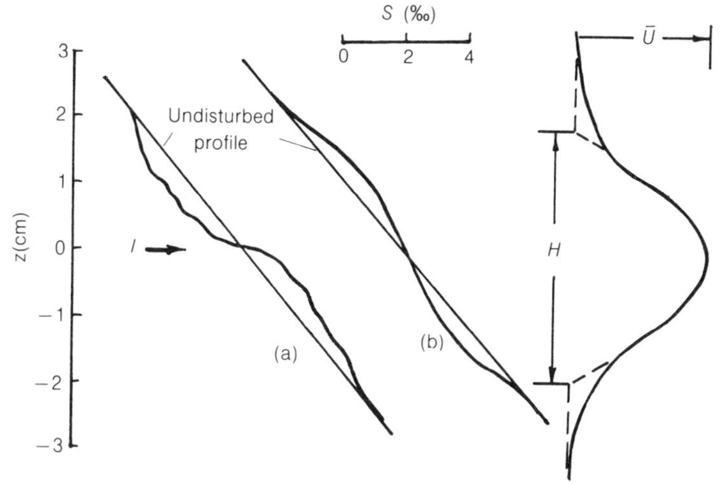

Figure 5.8 *Vertical profiles of horizontal velocity and salinity in a vortex dipole: (a) in a jet-like flow before the dipole is formed ($t < t_0$); (b) in the centre of the dipole after it has been formed ($t > t_0$). $\Delta t = 4$ s and Re = 530.*

formed are similar to the density distributions in the steady jet region and in the frontal region of the starting jet respectively (see Figure 5.5). Note also that the fluid in the vortex dipole is hardly mixed in the vertical direction.

Observations demonstrate that after the vortex dipole has been formed its shape is similar at any instant of time. For simplicity we shall consider a vortex dipole as a fluid disc of thickness H and diameter $D = 2R$, moving with a velocity $\bar{U} = \mathrm{d}L/\mathrm{d}t$, where L is the distance of the centre of the disc from the origin.

Under the standard hypothesis (e.g. Turner, 1964, 1973) that the influx q for a localized fluid volume is proportional to the mean velocity of the volume and its frontal area, the mass balance equation for an impulsive dipole is

$$\pi \frac{\mathrm{d}(HR^2)}{\mathrm{d}t} = q, \tag{5.2.1}$$

where $q = \pi \alpha H R \bar{U}$ and α is the entrainment coefficient. The momentum balance equation is

$$(1+k)\pi \frac{\mathrm{d}(HR^2\bar{U})}{\mathrm{d}t} = -F, \tag{5.2.2}$$

where $k = 1$ is the virtual mass coefficient and $F = c\nu R\bar{U}$ (similar to (5.1.10)) is the viscous drag force applied to the horizontal boundaries of the dipole. For solving (5.2.1) and (5.2.2) data on the dipole thickness H are needed. In general, H can depend on the kinematic momentum I, the buoyancy frequency $\mathcal{N}$, time t and viscosity ν. On dimensional grounds,

$$\frac{H\mathcal{N}^{1/4}}{I^{1/4}} = f(\tau_0, \tau_*), \tag{5.2.3}$$

where $\tau_0 = \mathcal{N}t$ and $\tau_* = \nu\mathcal{N}^{1/2}t/I^{1/2}$ are non-dimensional times. In experiments a strong jet is used to exert a large momentum for a short period of time on the fluid. Typical values of the governing parameters are $I \approx 10^2\,\mathrm{cm}^4\,\mathrm{s}^{-1}$, $\mathcal{N} \approx 1\,\mathrm{s}^{-1}$, $\Delta t \approx 5\,\mathrm{s}$ and $t \lesssim 10^2\,\mathrm{s}$. When $t > t_0$ a dipole is formed and $\tau_0 \gg 1$. In contrast, $\tau_* \ll 1$ when t is not too large, $t \lesssim I^{1/2}/\nu\mathcal{N}^{1/2}$. Thus, we assume complete similarity of the function $f(\tau_0, \tau_*)$ with respect to the parameters τ_0 and τ_*:

$$f(\tau_0 \gg 1, \tau_* \ll 1) = \text{const.}$$

Observations demonstrate that after the dipole has been formed $(t > t_0)$ its thickness H does not vary significantly with time; hence

$$f(\tau_0, \tau_*) = \text{const} = \gamma,$$
$$H = \gamma I^{1/4}\mathcal{N}^{-1/4}. \tag{5.2.4}$$

The results of measurements (Figure 5.9) for $t_0 \lesssim t \lesssim I^{1/2}/\nu\mathcal{N}^{1/2}$ give a mean value $\gamma \approx 1.4$.

Substitution for H and q allows integration of (5.2.1), to give

$$2R = \alpha L. \tag{5.2.5}$$

In the model the coefficient α is not known; its value must be obtained from experiments. The results of measurements (Figure 5.10) are in agreement with the estimate (5.2.5), and give a mean value $\alpha \approx 0.46$.

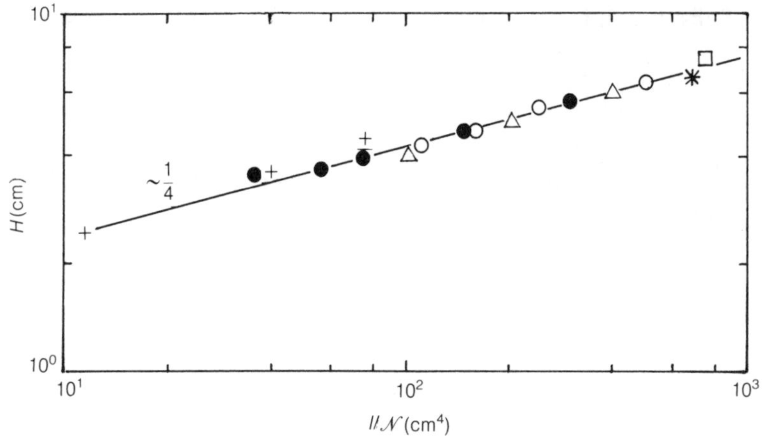

Figure 5.9 *Vortex dipole thickness H versus $I/\mathcal{N}$: the momentum I was altered by variation of q_* (0.5–3.0 cm^3 s^{-1}) and Δt (1–8 s). Symbols denote experimental points with different values of I (50–400 cm^4 s^{-1}) and $\mathcal{N}$ (0.5–2.5 s^{-1}). Each symbol represents a particular value of q_*. The solid line is (5.2.4) with $\gamma = 1.4$.*

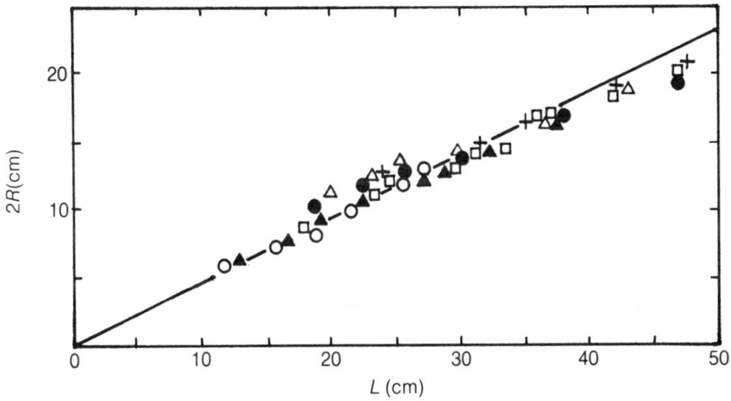

Figure 5.10 *Vortex dipole diameter $D = 2R$ for different distances L from the origin: symbols denote experimental points with different values of I (50–400 cm^4 s^{-1}) and $\mathcal{N}$ (0.5–2.5 s^{-1}). The solid line is (5.2.5) with $\alpha = 0.46$.*

Substitution for H, R and F allows integration of (5.2.2) to give

$$-L+\left(\frac{4I}{\alpha c v}\right)^{1/2}\operatorname{arctanh}\left[\left(\frac{\alpha c v}{4I}\right)^{1/2}L\right]=\frac{c v \mathcal{N}^{1/4}t}{\alpha(k+1)\gamma\pi I^{1/4}}. \quad (5.2.6)$$

In performing the integration, we have used the initial condition that for $t=0$ ($L=0$), momentum I is applied to the fluid. Expanding (5.2.6), we obtain the estimate

$$L\approx\left[\frac{12}{\alpha^2(1+k)\gamma\pi}\right]^{1/3}\mathcal{N}^{1/2}I^{1/12}t^{1/3}. \quad (5.2.7)$$

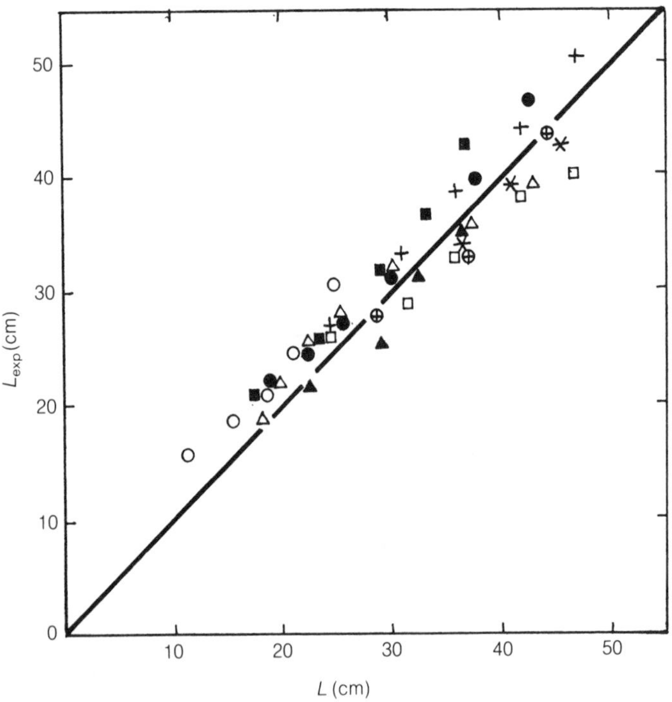

Figure 5.11 *Measured L_{exp} and calculated L (from (5.2.7) with $\alpha=0.46$ and $\gamma=1.4$) distances of the vortex dipole centre from the origin: each symbol denotes a different t (10–350 s) but the same I (50–400 $cm^4\,s^{-1}$) and $\mathcal{N}$ (0.5–2.5 s^{-1}).*

The results calculated from (5.2.7) are compared with experimental values of L in Figure 5.11.

Thus during the intermediate asymptotic stage (intermediate times) the dipole becomes self-similar, its diameter and distance from the origin increase with time as $t^{1/3}$, and the dipole thickness remains constant. Note that the global Reynolds number of the flow $\overline{Re} = 2R\bar{U}/\nu \propto t^{-1/3} \propto x^{-1}$ decreases with time. In contrast, the Richardson number $\overline{Ri} = H^2\mathcal{N}^2/\bar{U}^2 \propto t^{4/3} \propto x^4$ rapidly increases. Similar results can be obtained when the dipole moves between two homogeneous layers of different densities (Afanasyev, Voropayev and Filippov, 1988).

5.3 The flow into a dipole

The physical model of an impulsive dipole considered in section 5.2 gives estimates for the global properties of a dipole moving in a stratified fluid, but it does not consider the details of fluid motion within and around the dipole. To clarify the flow kinematics and to better understand the fluid particle motion into the dipole, one must perform some experiments with tracer particles. In Figure 5.12 two successive photographs of the flow show how the dipole moves through a cloud of small tracer particles. Polystyrene spheres of density approximately equal to the fluid density at the nozzle level were introduced into the resting stratified fluid before the start of the experiment. From successive photographs of the flow one can plot the positions of the particles and thus reconstruct their trajectories. Some trajectories plotted with respect to axes at rest are shown in Figure 5.13. The picture appears rather complex and chaotic, although the vortex dipole is obviously an ordered structure.

To elucidate the implicit order in the motion of particles, consider the flow with respect to axes connected with the moving dipole. We saw that the global properties of the dipole change similarly with time. Let us assume that the velocity field is also similar at different times. This assumption is equivalent in fact to the assumption of self-similarity of the flow. Choosing the radius of the dipole $R(t)$ as the current length scale, let us replot the particle trajectories in similarity coordinates $(\eta = r/R(t), \theta)$ with origin at the centre of the moving dipole. As a result, we obtain the regular picture (Figure 5.14b) of trajectories. For clarity the positions of only four tracer particles

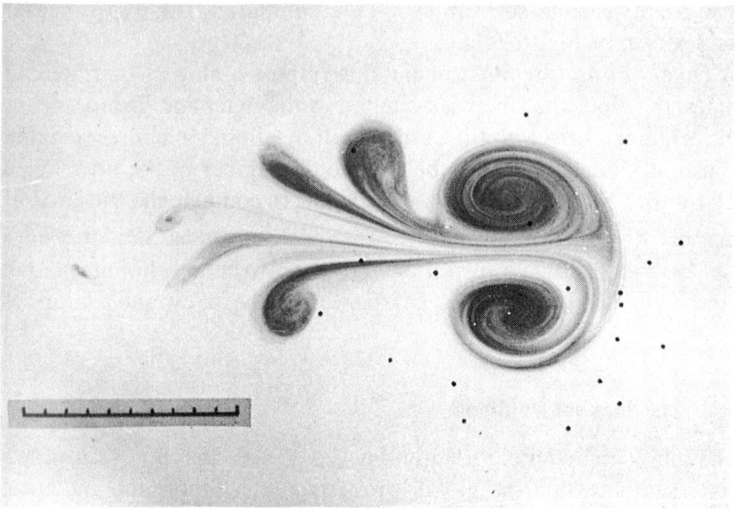

(a)

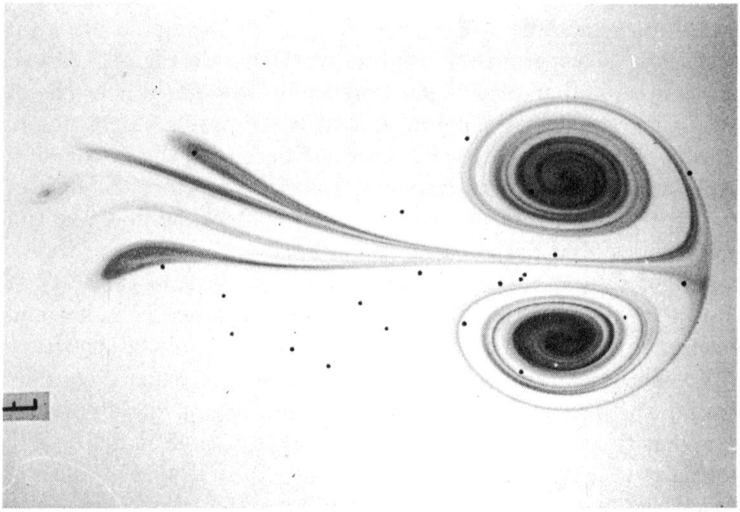

(b)

Figure 5.12 *Motion of tracer particles (polystyrene spheres) into an impulsive vortex dipole with* $I = 125\,\text{cm}^4\,\text{s}^{-1}$: *(a)* $t = 32\,\text{s}$; *(b) 125 s. The scale is in centimetres.*

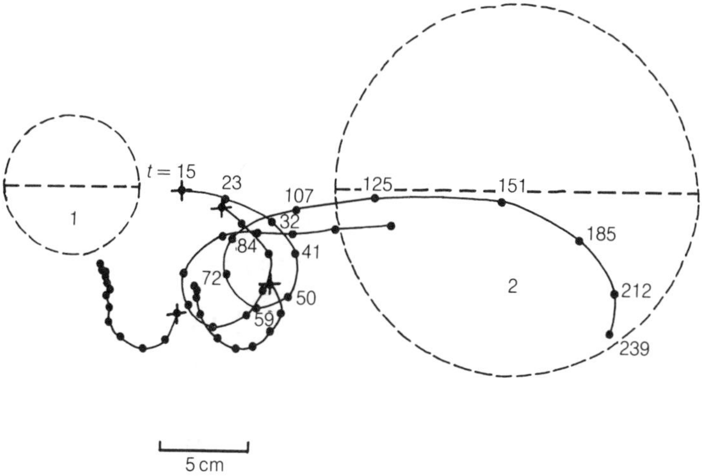

Figure 5.13 *Trajectories of four tracer particles plotted with respect to axes at rest, at time t in seconds from the beginning of the experiment. The initial (1) and final (2) positions of the dipole are shown by circles.*

(shown in Figure 5.13) are plotted. It can be seen that two of them enter into the inner region of the dipole and form typical spirals. Thus the flow can be separated into two parts by a circle of radius $\eta = 1$. The flow inside the circle is obviously vortical, while the flow outside resembles the irrotational flow around a rigid cylinder. The circular boundary turns out to be permeable to fluid particles.

In order to explain the experimental results shown in Figure 5.14(b), consider a simple kinematic model similar to that proposed by Turner (1964) for spherical thermals. The basic assumption is that the flow is instantaneously the same as for a steady vortex dipole of fixed size moving through a frictionless fluid. Thus the vorticity is concentrated in the dipole (inside a circle of radius R), while the outside flow is the potential flow around the cylinder. In this model the external ψ_e and internal ψ_i stream functions are given by (3.1.16). One must modify (3.1.16) to allow for the expansion in radius and decrease in translational velocity of the dipole. In the non-dimensional coordinates (η, θ) the corresponding velocity components can

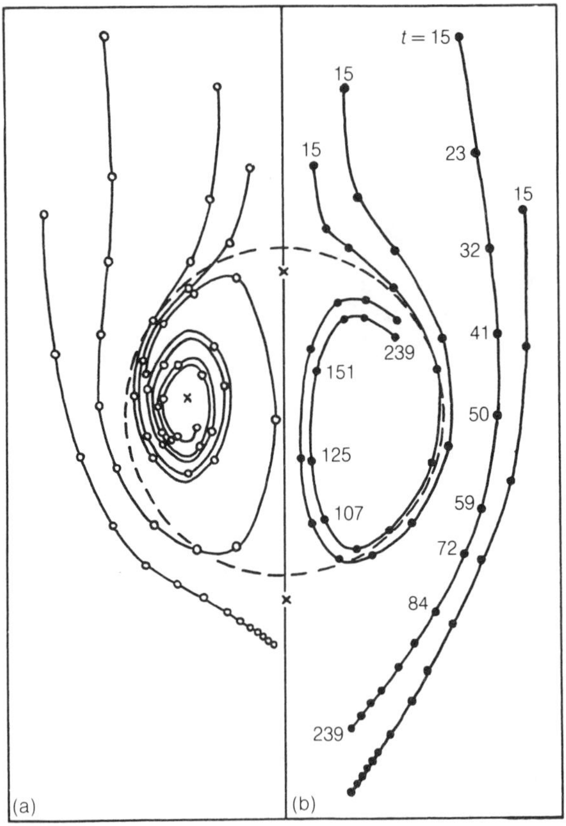

Figure 5.14 *Trajectories of four tracer particles (shown in Figure 5.13) in coordinates* ($\eta = r/R(t), \theta$)*: (a) calculations, (b) experiment. The time t is in seconds from the beginning of the experiment.*

be expressed in the form

$$u \equiv \frac{\mathrm{d}r}{\mathrm{d}\theta} = \eta \frac{\mathrm{d}R}{\mathrm{d}t} + R \frac{\mathrm{d}\eta}{\mathrm{d}t},$$

$$v \equiv \frac{1}{r} \frac{\mathrm{d}\theta}{\mathrm{d}t} = \frac{1}{\eta R} \frac{\mathrm{d}\theta}{\mathrm{d}t}.$$

On the other hand, u and v are related to a stream function:

$$u \equiv \frac{1}{r}\frac{d\psi}{d\theta} = \frac{1}{\eta R}\frac{d\psi}{d\theta},$$

$$v \equiv -\frac{d\psi}{dr} = -\frac{1}{r}\frac{d\psi}{d\eta}.$$

Substituting into the above relations the stream function in the form (3.1.16) and using the expressions for R and $\bar{U}$ obtained in section 5.2,

$$\bar{U} = \frac{2}{\alpha}\frac{dR}{dt}, \qquad R \propto t^{\beta}, \qquad \alpha \approx 0.45, \qquad \beta = \tfrac{1}{3},$$

after some algebra we obtain the equations of motion of fluid particles in the similarity coordinates (η, θ):

$$\left.\begin{aligned} \frac{d\eta}{dt} &= -\frac{2}{3\alpha t}\left[\tfrac{1}{2}\alpha\eta + \left(1 - \frac{1}{\eta^2}\right)\cos\theta\right], \\ \frac{d\theta}{dt} &= \frac{2}{3\alpha t\eta}\left(1 + \frac{1}{\eta^2}\right)\sin\theta \end{aligned}\right\} (\eta > 1) \qquad (5.3.1)$$

$$\left.\begin{aligned} \frac{d\eta}{dt} &= -\frac{2}{3\alpha t}\left[\tfrac{1}{2}\alpha\eta + \frac{2J_1(\alpha\eta)}{J_0(c)\eta}\cos\theta\right], \\ \frac{d\theta}{dt} &= \frac{2}{3\alpha t\eta}\frac{J_1(\alpha\eta)}{J_0(c)}\sin\theta \end{aligned}\right\} (\eta < 1). \qquad (5.3.2)$$

These equations can be integrated numerically. The integration may be started with a set of initial values (t_0, η_0, θ_0) that are the same as those in the experiment with the tracer particles. First the system (5.3.1), which represents the flow in the exterior region, should be integrated until $\eta = 1$. Then the motion for $\eta < 1$ may be followed using (5.3.2). For comparison, the calculated trajectories of four particles, corresponding to the experimental trajectories of four polystyrene spheres in Figure 5.14(b), are shown in Figure 5.14(a). The stagnation points where $\partial\psi/\partial\eta = \partial\psi/\partial\theta = 0$ are marked by crosses. The qualitative agreement is quite satisfactory, and the model correctly describes the main feature of the flow – its spiral character. Note that in the model the whole boundary of the dipole is permeable to the exterior fluid, while in the real flow the particles enter the dipole from its rear. This disadvantage of the model is a consequence of the discontinuity of the second derivatives of the stream function (3.1.16) at $\eta = 1$. In the real flow the distribution of vorticity is obviously smoothed by viscous diffusion, so that the discontinuity cannot occur.

5.4 A developing vortex quadrupole

In its archetypal form, a vortex quadrupole is formed in a viscous fluid under the action of a localized force dipole. The total momentum of the flow induced by this forcing is zero. The action of a localized force dipole can be modelled experimentally by a horizontally oscillating vertical cylinder (Figure 1.7). The action of a small oscillating cylinder is equivalent to the action of a point force dipole with intensity Q, given by (1.2.6).

The typical evolution of the flow induced by an oscillating cylinder in a thin layer of fluid can be observed in the sequence of plan-view photographs presented in Figure 5.15. The cylinder itself is not visible, since it points downwards into the fluid and performs a horizontal oscillating motion in a direction perpendicular to the axis of the supporting frame (Figure 1.7). At the start of the experiment some dark dye was introduced near the cylinder; most of this dye is seen to get carried away in the two symmetric vortex dipoles that are formed after the oscillation has started. The two vortex dipoles move away from the source along aligned straight trajectories; their size is observed to increase gradually, while their translational speed decreases.

A sketch of the developing vortex quadrupole is shown in Figure 5.16. The flow is planar, thus it can be described by just two global properties: the typical length L and the typical width D. The problem has no external length scale and the only scale is the distance r from the origin. To obtain estimates for D and L, let us again explore the idea of similarity and suppose that at the intermediate asymptotic stage the flow develops self-similarly. In this case, by taking the distance r as a typical length scale L of the flow, we can write

$$D = \alpha L. \tag{5.4.1}$$

By scaling the flow velocity field by some reference velocity scale U_0, the velocity $\bar{U}$ of the vortex front moving into the ambient

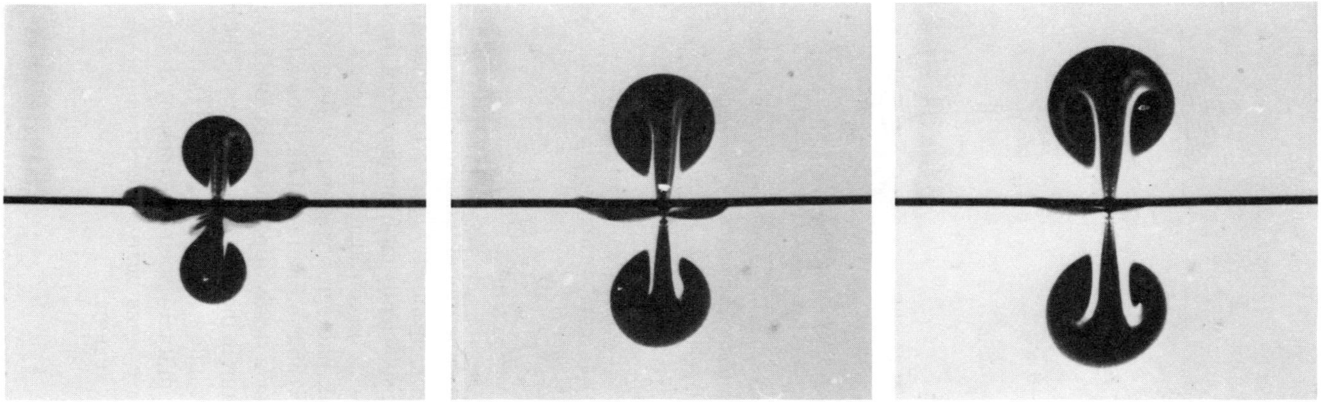

(a) (b) (c)

Figure 5.15 *Formation and subsequent evolution of a vortex quadrupole in a thin fluid layer. The flow is driven by a rapidly oscillating cylinder. This cylinder is fixed to the supporting frame, which is visible on the photographs as a dark line.* $f = 9\ Hz$, $\varepsilon = 0.14\ cm$, *and times* $t = 4.4\ s$ *(a),* $9.2\ s$ *(b) and* $13.2\ s$ *(c) after the forcing was started.*

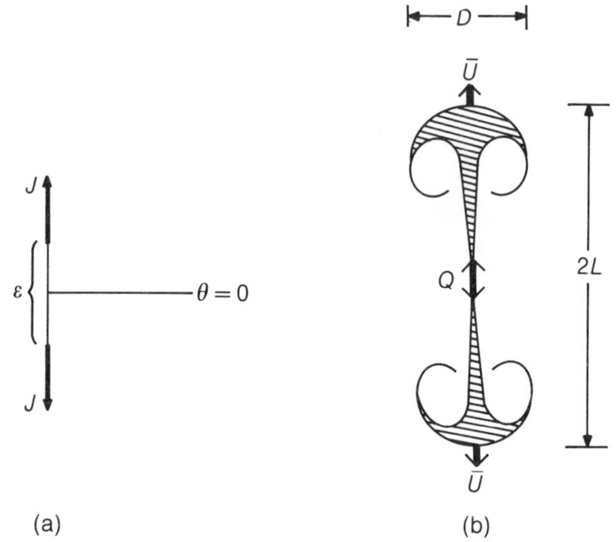

Figure 5.16 *Schematic drawing of the momentum sources (a) and the quadrupolar vortex (b) arising under the acting of two equal forces acting in opposite directions. The typical length L, width D and propagation velocity* $\bar{U}$ *(* $\equiv dL/dt$ *) of the evolving dyed flow structure are defined on the figure.*

quiescent fluid (Figure 5.16) can be estimated in a similar fashion as

$$\bar{U} \equiv \frac{dL}{dt} = \beta U_0, \tag{5.4.2}$$

where α and β are non-dimensional functions of forcing.

To estimate U_0, look again at Figure 5.15. Once the dipolar structures have moved sufficiently far from the origin the flow in the vicinity of the source can be considered to be steady. This can be observed on the streak photographs shown in Figure 5.17. When the vortex dipoles still visible in Figures 5.17(a, b) have left the source region, the flow takes on a steady appearance (Figure 5.17c), consisting of two outflow regions in the form of narrow jets in the oscillation direction and two wide inflow regions (Figure 3.17). In the flow field near the source the azimuthal velocity component is negligible and the tracer particles move along nearly straight radial lines, in both the inflow and outflow regions.

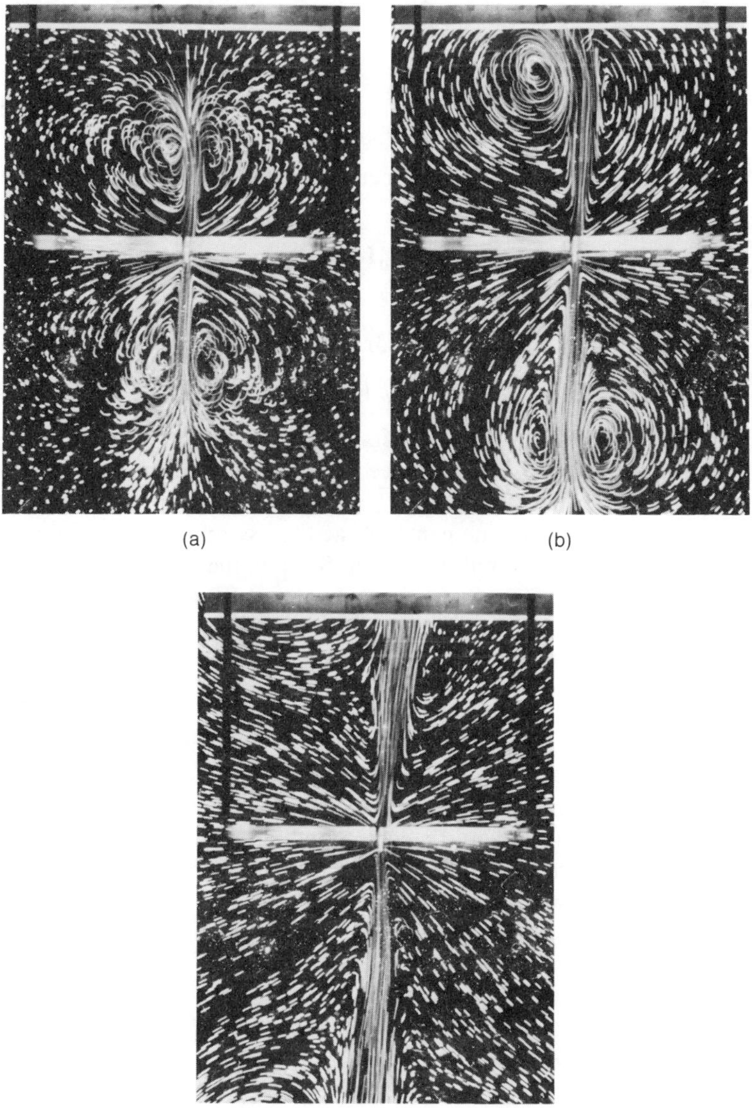

(a) (b) (c)

Figure 5.17 *Streak photographs showing the evolution of the vortex quadrupole induced by an oscillating cylinder: (a) $t=4$ s; (b) 10 s; (c) 15 s. The rectangular construction visible in the pictures is the supporting frame. $f=9$ Hz, $\varepsilon=0.14$ cm and the exposure time was 2 s.*

The solution (3.7.21) for the steady asymptotic gives an estimate for the reference velocity:

$$U_0 = Re\frac{\nu}{r} \quad \text{at} \quad r = L, \tag{5.4.3}$$

where Re is the Reynolds number of the flow, which is equal here to the non-dimensional forcing

$$Re = \frac{U_0 L}{\nu} = \frac{Q}{4\pi\nu^2}. \tag{5.4.4}$$

Combination of (5.4.2) and (5.4.3) yields

$$L = (2\beta\, Re\, \nu t)^{1/2}. \tag{5.4.5}$$

Thus, if the flow develops self-similarly, only two non-dimensional functions $\alpha(Re)$ and $\beta(Re)$ are needed to estimate the global characteristics of the flow: D, L and $\bar{U}$. Using the solution (3.5.28) for a starting quadrupole, one can look at the asymptotic behaviour of $\alpha(Re)$ and $\beta(Re)$ for small Re values. Since in the experiments α and β are determined from the measured values of D and L in the dyed structure, one has to analyse the equations of motion for marked particles

$$\frac{\mathrm{d}r}{\mathrm{d}t} = u, \quad \frac{\mathrm{d}\theta}{\mathrm{d}t} = \frac{v}{r}, \tag{5.4.6}$$

to obtain the contour of the dyed region in the developing flow. The velocity components u and v corresponding to the stream function (3.5.28) are written as

$$u = \frac{1}{r}\frac{\partial\psi}{\partial\theta} = -Re\frac{\nu}{r}\eta^{-2}\{1 - \exp(-\eta^2)\}\cos 2\theta, \tag{5.4.7a}$$

$$v = -\frac{\partial\psi}{\partial r} = -Re\frac{\nu}{r}\{\eta^{-2} - (1 + \eta^{-2})\exp(-\eta^2)\}\sin 2\theta. \tag{5.4.7b}$$

Thus the equations of motion (5.4.6) can be presented in similarity coordinates ($\eta = r/(2\sqrt{\nu t}), \theta$):

$$\frac{\mathrm{d}\eta}{\mathrm{d}\tau} = -\frac{Re}{4}\eta^{-3}\{1 - \exp(-\eta^{-2})\}\cos 2\theta - \tfrac{1}{2}\eta, \tag{5.4.8a}$$

$$\frac{d\theta}{d\tau} = -\frac{Re}{4}\eta^{-2}\{\eta^{-2} - (1+\eta^{-2})\exp(-\eta^2)\}\sin 2\theta, \qquad (5.4.8b)$$

where $\tau = \log t$.

The points where the non-dimensional velocities are equal to zero,

$$\frac{d\eta}{d\tau} = 0, \quad \frac{d\theta}{d\tau} = 0,$$

are critical stagnant points in the similarity coordinate frame. Two symmetrical points $(\eta = \eta_0, \theta = \pm\pi/2)$ give the positions of the leading dyed fronts of the flow. Here the value η_0 is given by the relation

$$\frac{Re}{2}\eta_0^{-3}\{1 - \exp(-\eta_0^2)\} = \eta_0, \qquad (5.4.9)$$

obtained from (5.4.8) at $\theta = \pm\pi/2$. Using (5.4.7), we obtain the following for the propagation velocity $\bar{U}$ of the leading fronts

$$\bar{U} = u\left(\eta = \eta_0, \theta = \pm\frac{\pi}{2}, t\right) = \left(\frac{\nu}{t}\right)^{1/2}\eta_0 \qquad (5.4.10)$$

or in equivalent form

$$\bar{U} = 2\frac{\nu}{r}\eta_0^2.$$

Thus we have obtained the value of the propagation velocity of the front formed by the dyed particles which were initially in the vicinity of the source. Scaling $\bar{U}$ by U_0 given by (5.4.3) gives

$$\beta = \frac{\bar{U}}{U_0} = \frac{2}{Re}\eta_0^2. \qquad (5.4.11)$$

Solving (5.4.9) we obtain an estimate for $\beta(Re)$ which is shown graphically in Figure 5.18 for small Re values. The maximum length L of the dyed structure is then given by

$$L = \int_0^t \bar{U}\,dt = 2(\nu t)^{1/2}\eta_0. \qquad (5.4.12)$$

The maximum width D of the flow is determined by calculating the positions of the points at which the velocity component perpendicular

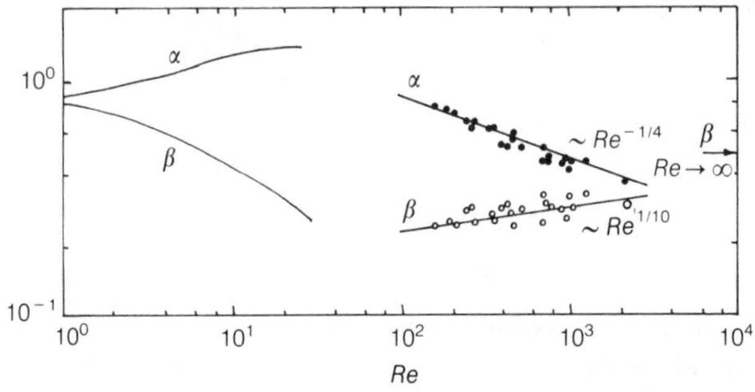

Figure 5.18 *The behaviour of α(Re) and β(Re) for a wide range of Re values. The solid lines represent theoretical estimates using (5.4.11) and (5.4.14) (for small Re values) and approximations using (5.4.15) and (5.4.16) (for large Re values) of the experimental data shown by points. The transition region at moderate Re values remains unstudied. The experimental parameters for each point are given in Voropayev, Afanasyev and van Heijst (1993a).*

to the axis of the flow changes its sign. This gives the condition

$$\frac{d}{d\tau}(\eta \cos \theta) = 0,$$

which describes the curve

$$\theta = \sin^{-1}\left[\frac{2\eta^4 Re^{-1} + 1 - e^{-\eta^2}}{2(2(1 - e^{-\eta^2}) - \eta^2 e^{-\eta^2})}\right]^{1/2}. \tag{5.4.13}$$

The maximum distance between symmetrical points $(\eta^\circ, \theta^\circ)$ and $(\eta^\circ, \pi - \theta^\circ)$ on this curve gives the value of the width of the dyed structure

$$D = 2(\nu t)^{1/2}\,(2\eta^\circ \cos \theta^\circ).$$

The values of

$$\alpha(Re) = \frac{D}{L} = 2\frac{\eta^\circ}{\eta_0}\cos \theta^\circ \tag{5.4.14}$$

calculated from (5.4.13) are also shown in Figure 5.18.

The estimates obtained for the functions $\alpha(Re)$ and $\beta(Re)$ are valid only for small Re values. From the general case one might expect that when nonlinear effects become significant, $\alpha(Re)$ decreases as

Re increases with the limit $\alpha(Re) \to 0$ at $Re \to \infty$ (e.g. Batchelor, 1967) while $\beta(Re) \to \frac{1}{2}$ at $Re \to \infty$ (Stern and Voropayev, 1984). To reveal the behaviour of α and β for real flows at large *Re* values one has first to verify the validity of (5.4.1) and (5.4.5) for *D* and *L* in experiments.

Typical experimental data for the width *D* of the quadrupolar structure and the distance *L* from the origin to the front of the dyed structure are shown in Figure 5.19. To demonstrate that the size of the working container does not significantly influence the experimental results, two sets of experimental data, obtained in small and large containers at approximately the same *Re* values, are shown together in this figure. It can be seen that the experimental data are in good agreement with the linear relationship (5.4.1). Mean values of the coefficient α have been determined for every experiment and the results are plotted in Figure 5.18 as a function of the Reynolds number. The straight line in this figure represents an approximation to the data by a conjectured power law of the form

$$\alpha = c_1 \, Re^{-1/4}, \tag{5.4.15}$$

where $c_1 = 2.5$. In order to verify (5.4.5), the dependence of *L* on time *t* has to be measured experimentally. Some typical results are presented in Figure 5.20, and good agreement with a power law

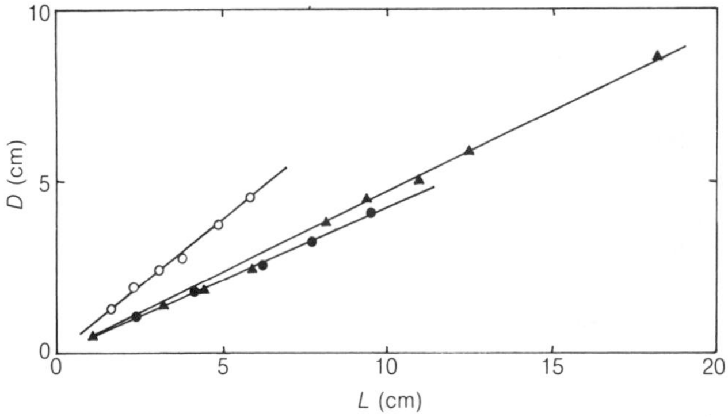

Figure 5.19 *Typical measured values of the width D of the vortex quadrupole for different distances L from the origin. Experimental parameters:* ○ $d = 0.12$ *cm,* $\varepsilon = 0.1$ *cm,* $f = 9$ *Hz,* $Re = 190$; ● $d = 0.16$ *cm,* $\varepsilon = 0.12$ *cm,* $f = 18$ *Hz,* $Re = 980$; ▲ $d = 0.24$ *cm,* $\varepsilon = 0.1$ *cm,* $f = 17$ *Hz,* $Re = 935$; ○, ● *small tank;* ▲ *large tank.*

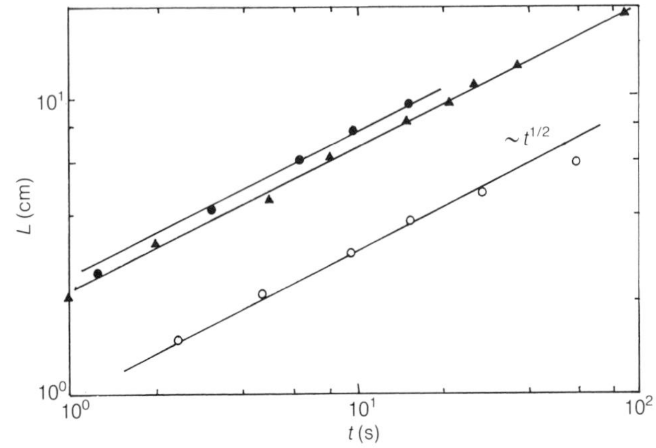

Figure 5.20 *Measured values of the length L for different times t. The experimental parameters are the same as in Figure 5.19.*

$L \propto t^{1/2}$ can be observed. The proportionality coefficients can be estimated from these data, and this yields the values of the function $\beta(Re)$ in (5.4.5). The results are presented graphically in Figure 5.18 (together with those for α) as a function of the Reynolds number and are approximated by the power law

$$\beta = c_2 Re^{1/10}, \tag{5.4.16}$$

where $c_2 = 0.14$.

It is not possible to compare directly the data for the functions α and β with the corresponding data for a developing vortex dipole (Figure 5.6), since the forcing and the definition of β are different for the dipolar and quadrupolar flows. In the case of a quadrupole the linear solution is used to estimate the velocity scale U_0, while in the case of the dipole the nonlinear solution (2.3.29) is used for a similar purpose. Nevertheless, the functions α and β for these two types of flow are qualitatively similar: in both cases the function α shows a rapid decrease with *Re* (this means that the flow becomes narrower for stronger forcing) and the function β shows only a slight increase with *Re* at large *Re* values.

Thus for a broad range of *Re* values the quadrupolar flow develops in similar manner and its global characteristics are determined by two non-dimensional functions $\alpha(Re)$ and $\beta(Re)$ shown in Figure 5.18

(only a narrow transition region, $Re = 10-100$, remains unstudied). These results are used below to interpret experimental data on the evolution of an array of quadrupoles.

5.5 An array of quadrupoles

Now consider a more complicated forcing: a system of force dipoles generating an array of vortex quadrupoles, developing in a direction

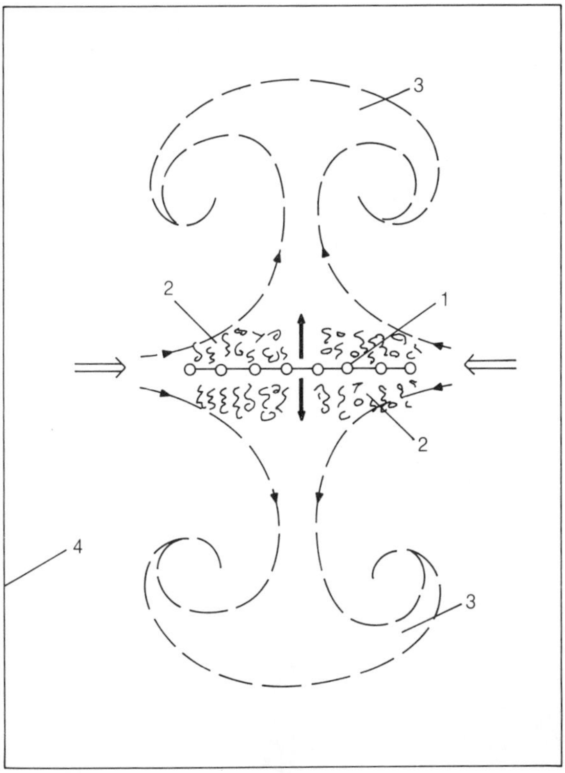

Figure 5.21 *Schematic drawing of the flow generated by an oscillating grid with the tank side walls far from the ends of the grid. 1 = the grid; 2 = the region of small-scale vortical motion near the grid; 3 = the induced large-scale quadrupolar flow, which is governed by the total forcing $Q_0 = Q\ D_0/d_0$ of the grid; 4 = the container. The arrows indicate the inflow of ambient fluid near the ends of the grid.*

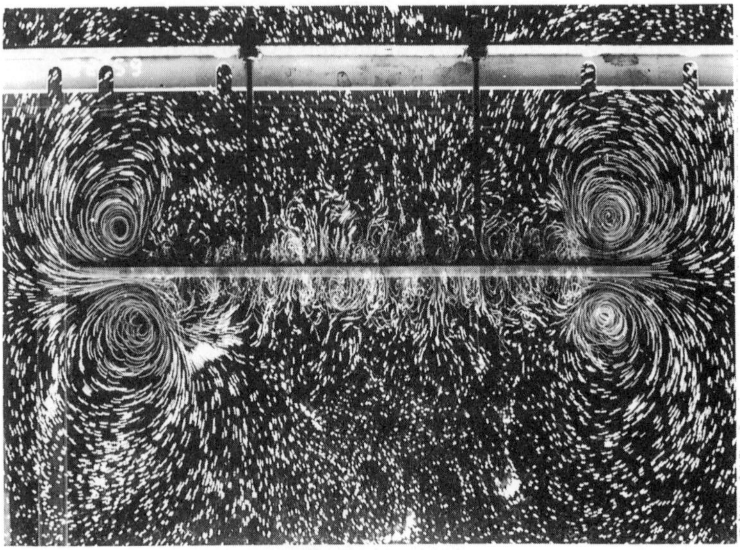

(a)

(b)

(c)

Figure 5.22 Sequence of streak photographs of the flow generated by the oscillating grid with the tank walls far ($\simeq 8$ cm) from the grid ends. Experimental parameters: $d = 0.07$ cm, $d_0 = 1$ cm, $\varepsilon = 0.13$ cm, $f = 18$ Hz, exposure time was 2 s. The rectangular construction visible in the upper parts of the photographs is the supporting frame to which the grid was attached.

normal to the line on which the force dipoles are distributed. To model this kind of forcing, a linear grid of vertical rods oscillating horizontally in a thin upper layer of fluid is used.

First we will consider the flow induced by the grid when its ends are far from the side walls of the container (Figure 5.21). After the grid oscillations are started, each oscillating rod acts on the fluid as a localized force dipole with intensity Q per unit of immersed rod length, given by (1.2.6). Thus the action of the oscillating grid can be taken to be a distributed force dipole of total intensity $Q_0 = QD_0/d_0$ (D_0 is the length of the grid and d_0 is the spacing between rods), so that at large distances the flow must be similar to that produced by a single force dipole of intensity Q_0. In accordance with (3.5.28) the far-field flow has a quadrupole character with the stream function given by

$$\psi_0 = \lim_{\eta \to \infty} \psi = -\frac{Q_0 t}{2\pi r^2} \sin 2\theta.$$

In a incompressible fluid this large-scale motion is induced instantaneously by the action of pressure forces. This motion is very strong (in the experiments $D_0/d_0 = 10$–40, hence, $Q_0 \gg Q$) and it prevails over the flows induced by the individual rods. As a result, the vorticity produced by each rod is overwhelmed by this motion, leading to the formation of four large vortices (Figure 5.22a). The flow in a confined container becomes quasi-steady with time: the irregular small-scale motions are localized in a small region near the grid, while the four large vortices are at some distance from it. Although these large cells drift slightly about a mean position, the four-cell pattern is fairly stable (Figures 5.22b, c). This flow is very similar qualitatively to the quasi-steady streaming induced by an oscillating

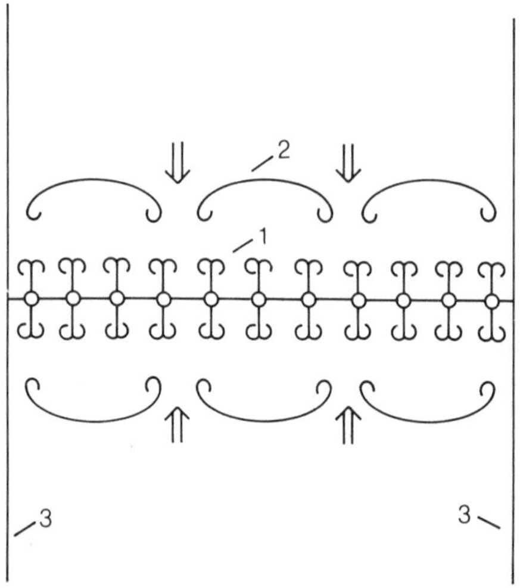

Figure 5.23 *Schematic drawing of the flow generated by an oscillating grid stretched across the width of the channel. 1 = primary quadrupoles; 2 = the region where the primary structures intensively interact, forming larger secondary vortex structures; 3 = the channel walls. The arrows indicate the entrainment of ambient fluid.*

cylinder in a small tank (see Schlichting, 1955 and Figure 31 in Van Dyke, 1982). Thus the array of localized forced dipoles acting on the fluid in the geometry under consideration induces an intensive large-scale quadrupolar circulating motion.

To eliminate this large-scale motion, the grid was placed in the middle of the channel such that it spanned the entire width of the channel (Figure 5.23). In this geometry the grid does not produce any significant large-scale motion and after the grid oscillations are started its action can be described as a linear arrangement of force dipoles. This forcing gives rise to a linear array of vortex quadrupoles, as can clearly be seen in Figure 5.24a. In the initial stage of the experiment (the first 1–2 s), when the typical width D of the primary vortex structures is still less than the rod spacing d_0, the quadrupoles develop independently (Figure 5.24a). As time progresses, D gradually increases and when D becomes comparable to the spacing distance d_0 the quadrupoles start to interact (Figure 5.24b). The explicit order is lost and the flow is chaotic or turbulent (Figures 5.24c, d). The short initial stage is over and a prolonged (many tens of seconds) intermediate stage begins. At this stage irregular two-dimensional motions are confined to a limited region with well defined sharp boundaries. The boundaries propagate in opposite directions away from the grid into the irrotational ambient fluid. It is the limited size of the turbulent region, i.e. its well defined width H (Figure 5.24) that provides an easily measurable parameter that can be used to quantify the global characteristics of the flow.

Visual observations revealed a rather complex picture of small-scale motions in the turbulent region. In the vicinity of the grid small but intensive quadrupoles are generated. The small-scale primary vortices, generated on both sides of the grid, are not seen to interact directly with their counterparts on the opposite side of the grid (in fact the flows on the opposite sides of the grid influence one another by pressure forces, as in the case of single quadrupole). Observations show that these small-scale primary vortices directly interact mostly with vortices generated on the same side of the grid by the nearest rods. The neighbouring primary dipoles merge, forming a larger secondary dipolar structure that detaches from the source of motion and moves away from the grid. At the same time a new small primary structure grows in the vicinity of the rod and the process is repeated. The size of the secondary structures gradually increases due to the

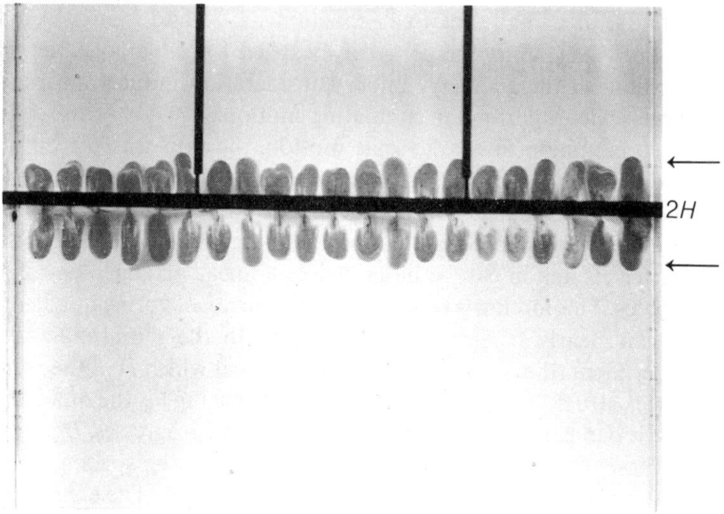

(a)

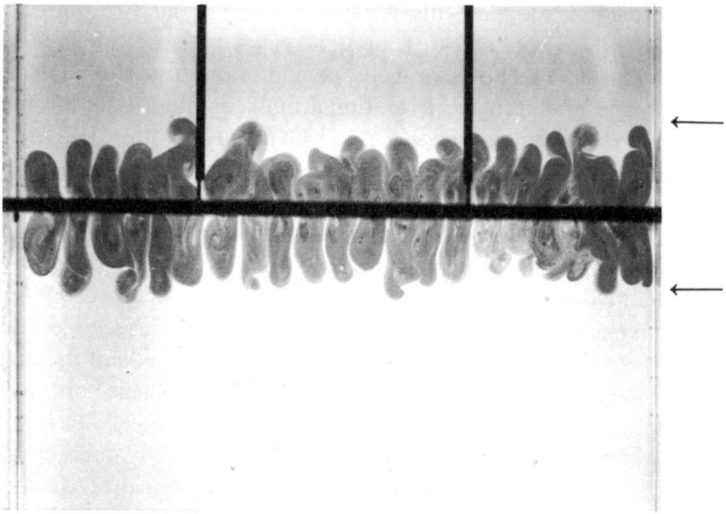

(b)

Figure 5.24 *Sequence of photographs showing the evolution of the flow generated by a horizontally oscillating linear grid of vertical rods in a channel. The forcing was started at $t = 0$, and the photographs were taken at (a) $t = 1.3$ s; (b) 3.2 s (c) 5.5 s and (d) 19 s. Experimental parameters: $d = 0.07$ cm,*

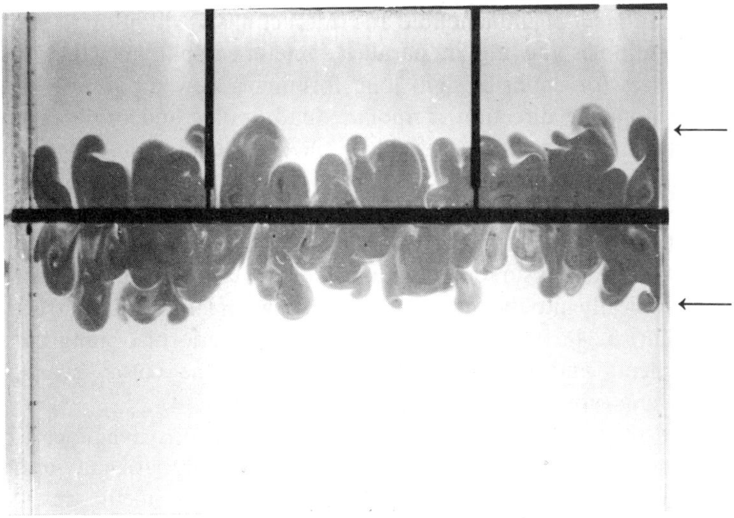

(c)

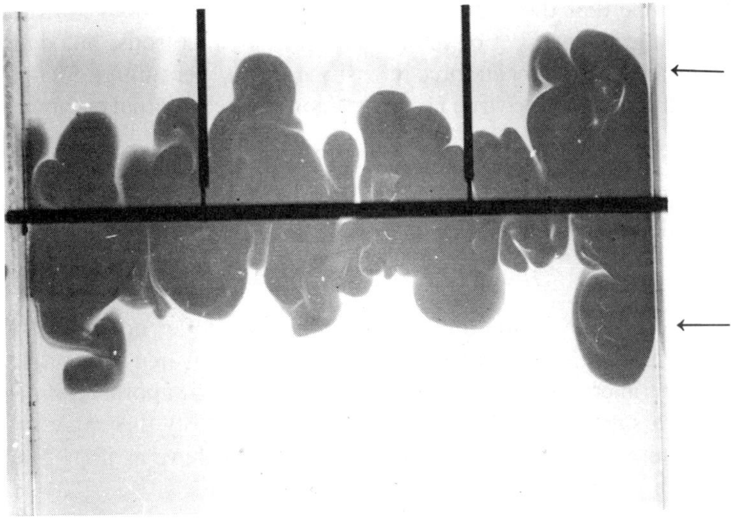

(d)

$d_0 = 1$ cm, $D_0 = 20$ cm, $\varepsilon = 0.13$ cm, $f = 18$ Hz, $Re = 530$. Just before the start of the experiment the fluid in the vicinity of the grid was dyed. The arrows indicate the mean position of the vortex fronts.

entrainment of ambient fluid and interactions. In addition to pairing, two dipoles moving in parallel sometimes split and their inner vortices (of opposite sign) join, forming a new dipole moving in the opposite direction. Tripolar, quadrupolar and more complex structures are sometimes visible in the turbulent region, but the most common structure is dipolar. The interaction/pairing process is repeated continuously, leading to the formation of even larger tertiary structures and causing the vorticity front to propagate into the irrotational surrounding fluid. The irrotational fluid is entrained periodically into the turbulent region between large leading structures (Figure 5.24). It spreads in this region in the form of long narrow filaments and is gradually entrained into the vortex structures, forming typical spirals.

Instead of focusing on small-scale motions, the dynamics of the turbulent region can also be described by its global behaviour, e.g. by the width H of the turbulent region. In experiments this width is determined by measuring the mean position of the front of the dyed region relative to the grid. To derive an estimate of $H(t)$ let us assume that the width H of the turbulent region does not depend on the intensity Q of the forcing due to single rods, but depends instead on the mean force intensity $g = Q/d_0$ per unit grid length. It is then possible to write H as a function of three external parameters:

$$H = H(g, \nu, t). \tag{5.5.1}$$

The dimensions of the quantity g are $L^3 T^{-2}$. By applying a dimensional analysis, equation (5.5.1) can be written as

$$H = g^{1/3} t^{2/3} \varphi_1(\xi), \text{ with } \xi = g t^{1/2} \nu^{-3/2}, \tag{5.5.2}$$

where φ_1 is a non-dimensional function of the non-dimensional variable ξ. The typical evolution of $H(t)$, as measured in two different experiments, is shown in Figure 5.25. In the first approximation, the data suggest that $H \propto t^n$, with $n \simeq 1/2$. To satisfy this experimental result, the function $\varphi_1(\xi)$ in (5.5.2) should behave as if $\varphi_1(\xi) \propto \xi^m$, with $m = 2n - 4/3$. This would imply that

$$H = \text{const} \times g^{2n-1} \nu^{2-3n} t^n,$$

which is clearly incorrect. For example, for $n = 1/2$ this gives $H = \text{const}\,(\nu t)^{1/2}$, which is obviously independent of the main governing parameter g; this result also disagrees with the observations

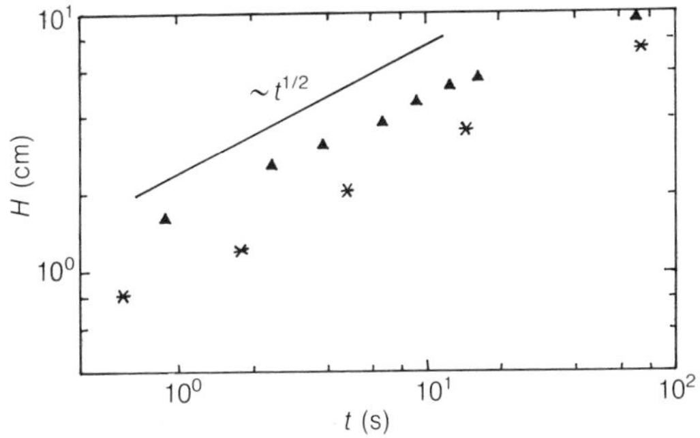

Figure 5.25 *Measured values of the width H of the turbulent region as a function of time t for two different experiments with Re = 820 (∗) and Re = 530 (▲).*

shown in Figure 5.25. The choice $n < 1/2$, as suggested by some of the data sets, would lead to $H \propto g^{-\delta}$, with $\delta = 1 - 2n > 0$, which is even more erroneous. To obtain a realistic estimate of H, we should return to the full set of governing parameters. In general, the width H can be written as a function of five parameters:

$$H = H(Q, D_0, d_0 \nu, t). \tag{5.5.3}$$

Two of these parameters have independent dimensions: hence, in non-dimensional form the number of parameters can be reduced from five to three. However, this is still a large number, and the chances of predicting the proper dependence are very small. Therefore it is useful, whenever possible, to take certain physical arguments into account. For this purpose consider the initial stage of the flow's evolution, when the primary vortex structures develop independently, without any direct mutual interaction. The typical length L and width D of the individual vortex quadrupoles increase with time according to (5.4.1) and (5.4.5) with α and β given by (5.4.15) and (5.4.16). By using these length scales dependence (5.5.3) can be written in non-dimensional form as

$$H/L = \varphi(Re, D_0/D, d_0/D). \tag{5.5.4}$$

Note that both L and D are introduced in (5.5.4) for the sake of convenience: since D and L are related through the known function $\alpha(Re)$, formally only one length scale is used in (5.5.4). Since in the experiments $Re \gg 1$ and $D_0/D \gg 1$ it is assumed that the function φ possesses complete similarity with respect to the arguments Re and D_0/D, i.e.

$$\varphi(Re \gg 1, D_0/D \gg 1, d_0/D) = \varphi_0(d_0/D).$$

Hence, the estimate for H is

$$H = L\varphi_0(d_0/D). \tag{5.5.5}$$

To determine the non-dimensional function $\varphi_0(d_0/D)$ experimentally, in all experiments the argument d_0/D as well as the length scale L have to be normalized at some transitional point, $d_0/D = 1$, between the initial and intermediate-asymptotic stages. With the help of (5.4.1), (5.4.5), (5.4.15) and (5.4.16) we obtain an estimate of the transition time

$$t_* = \frac{d_0^2}{2c_1^2 c_2 \, Re^{3/5}}. \tag{5.5.6}$$

By introduction of a non-dimensional time $\tau = t/t_*$, one can cast (5.5.5) in normalized form, as

$$H_0 = \frac{c_1 H}{d_0 \, Re^{1/4}} = \tau^{1/2} \varphi_0(d_0/D), \tag{5.5.7}$$

where $\tau = 1$ at $d_0/D = 1$, $c_1 = 2.5$ and $c_2 = 0.14$ in accordance with (5.4.15) and (5.4.16).

At $\tau \leqslant 1$ ($t \lesssim t_*$) the primary vortices do not interact significantly and develop independently. Hence, one may expect that at the initial stage the width H of the vortex region is equal to the length L of single structures: $H = L$; this gives $\varphi_0 = 1$ for $\tau \lesssim 1$. The experimental data obtained at small times ($t \lesssim t_*$) are plotted in the non-dimensional form of (5.5.7) in Figure 5.26. The solid line represents estimate (5.5.7) with $\varphi_0 = 1$. The best fit through the experimental data gives

$$\varphi_0 = c_0 = 0.95 \text{ for } \tau \lesssim 1, \tag{5.5.8}$$

which is in good agreement with the estimate $\varphi_0 = 1$.

At $\tau > 1$ the initial stage is over, the primary vortices start to interact and the intermediate stage begins. At this stage the interactions play an essential role, the motion is turbulent and φ_0 in (5.5.7) may differ from (5.5.8).

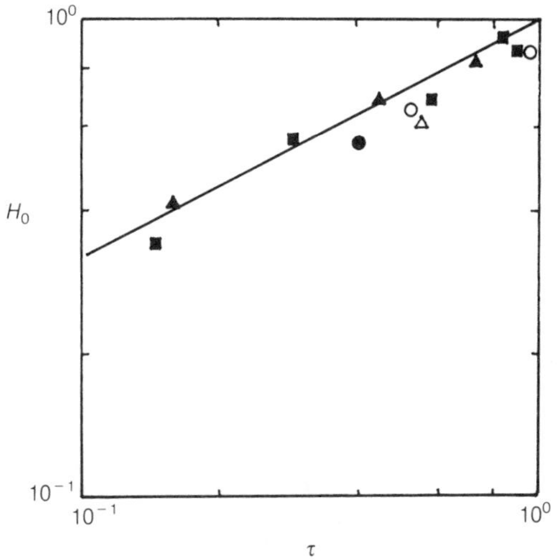

Figure 5.26 *Experimental data representing the normalized non-dimensional width H_0 of the vortex region as a function of the normalized time τ for the initial stage ($\tau \leqslant 1$). The solid line represents dependence (5.5.7) with an estimated value of function φ_0 ($\varphi_0 = 1$). Different symbols indicate different Re values in the range Re = 225–740.*

To compare the experimental data obtained at this stage ($\tau > 1$) with (5.5.7), a small correction ($t + \Delta t$) was introduced into the time origin. This was done because the origin of time in the experiment does not coincide with the origin of the intermediate stage as $H \to 0$. (Note that at $\tau > 1$ the explicit order, clearly visible during the initial stage in Figure 5.24a, is lost and the motion is turbulent over the entire width H of the vortex region; only in the vicinity of the source are distinct correlations with the grid motion observed.) The values of Δt were obtained by extrapolating the width H to zero on curves $H \propto t^{1/2}$. The measurements yield values of Δt in the range $\Delta t =$ 0–1 s. These small corrections were introduced into all the experimental data obtained for $\tau > 1$ and the data were replotted according to the non-dimensional form (5.5.7). The result is presented in Figure 5.27. The solid line in the graph represents estimate (5.5.7) with $\varphi_0 = \text{const}$. The best fit through the experimental points gives

$$\varphi_0 = c = 0.6 \text{ for } \tau > 1. \tag{5.5.9}$$

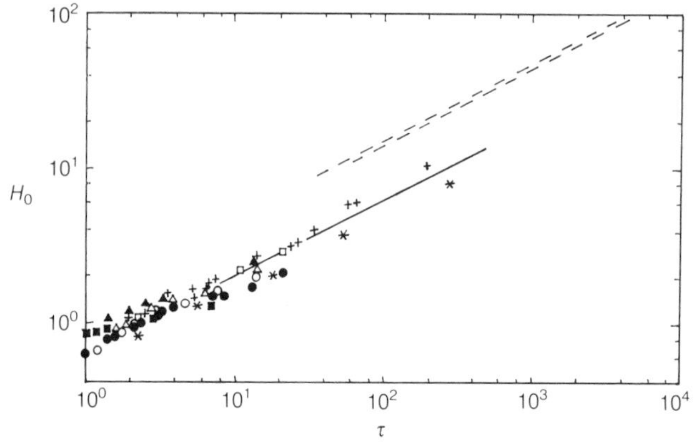

Figure 5.27 *Experimental data representing the non-dimensional width H_0 of the turbulent region as a function of the normalized time τ for the intermediate stage ($\tau > 1$). The solid line represents the estimate using (5.5.7) with $\varphi_0 = 0.6$. Different symbols indicate different Re values in the range Re = 225–820. The dashed lines represent the data obtained by Dickinson and Long (1978) in experiments with a plane oscillating grid. The experimental parameters for each point are given in Voropayev, Afanasyev and van Heijst (1993).*

Thus the function $\varphi_0(d_0/D)$ in (5.5.7) in the first approximation can be represented by a step-like function given by (5.5.8) and (5.5.9).

It can be concluded that the main governing parameter of the two-dimensional chaotic flow induced by the oscillating grid is the intensity of non-dimensional forcing, as in the case of a single quadrupole. The external length scale (a rod spacing of d_0) does not enter directly into the dependence $H(t)$, but is introduced indirectly via the transition time t_*. It appears that the physical arguments and the scaling analysis used to explain the propagation dynamics of a two-dimensional turbulent region are general enough to be applied to three-dimensional geometry as well.

When a planar horizontal grid oscillates vertically in a deep container of homogeneous fluid, it produces a horizontal layer of intense irregular three-dimensional motions. This turbulent region has a sharp boundary which gradually propagates into the surrounding irrotational fluid, thus giving rise to a deepening turbulent layer. After pioneering experiments by Rouse and Dodu (1955) and Cromwell (1960), this experimental arrangement was widely used by many authors to study different aspects of three-dimensional shear-free turbulence. For example, this forcing mechanism was applied to study the general nature of such turbulent flows (e.g. Thompson and Turner, 1975; Hopfinger and Toly, 1976) and the deepening of a turbulent layer in a homogeneous fluid (e.g. Dickinson and Long, 1978; Voropayev *et al.*, 1980). The forcing of the stirring grid was also used in various laboratory studies of mixing processes in a stratified fluid (e.g. Turner, 1968; Linden, 1975; Fernando, 1991).

This method was also used in the study of more complex phenomena such as stabilization/destabilization of turbulence by locally supplying a buoyancy flux through heating or by injection of a lighter or denser fluid near the planar oscillating grid (see, for example, Kantha and Long, 1980; Hopfinger and Linden, 1982; Benilov, Voropayev and Zhmur, 1983). The results of numerous experiments on three-dimensional oscillating-grid turbulence have stimulated theoretical investigations (e.g. Kraus and Turner, 1967; Long, 1978; Barenblatt and Voropayev, 1983), and appear to be relevant for a variety of flow situations, in particular for geophysical flows (see, for example, Turner, 1973; Kraus, 1977). Despite fundamental progress in this direction, one important question remains: what is the main governing parameter that characterizes the intensity of the forcing produced by the planar oscillating grid on the fluid? In general, this parameter is a function of the amplitude and frequency of oscillations, the geometry of the grid and the fluid's properties. Obviously, it is not the flux of linear (or angular) momentum, which is (averaged over the oscillation period) equal to zero. An attempt to take the flux of kinetic energy as the main governing parameter was not successful because this does not remain constant as the flow evolves (Voropayev *et al.*, 1980; Barenblatt and Voropayev, 1983; Benilov, Voropayev and Zhmur, 1983).

To demonstrate that the analysis used above for two-dimensional geometry can also be applied to the three-dimensional case, consider results reported by Dickinson and Long (1978). In their experiments a plane horizontal grid fabricated from a woven copper screen ($d = 0.033$ cm, $d_0 = 0.155$ cm) was used to generate three-dimensional turbulence. This grid was oscillated vertically in a deep tank and it produced a horizontal turbulent layer of increasing depth $H(t)$. The amplitude of oscillations was fixed, $\varepsilon = 0.475$ cm, and the frequency of oscillations was changed in different experiments in the range

$f = 6.3–12.0$ Hz. The transition time t_*, given by (5.5.6), for these parameters is very short ($t_* \simeq 0.02$ s) and only the intermediate stage was studied. The measurements obtained in these experiments are presented in the non-dimensional form (5.5.7) in Figure 5.27 and are shown by the two dashed lines between which points obtained in other experiments with different frequencies are situated. It is clear that the experimental data obtained in the different experiments practically lay on a single line, in agreement with dependence (5.5.7). This gives

$$\varphi_0 = c_3 = 1.3 \quad \text{for} \quad \tau > 1 \tag{5.5.10}$$

in the case of three-dimensional geometry. The difference in the values of φ_0 for two- and three-dimensional geometry is not surprising. In the latter case the motion can be considered as planar only in the close vicinity of the grid (at least, in the boundary layer near the rods) where no-slip conditions and viscous effects determine the amplitude of forcing, as given by (1.2.6). At greater distances the motion is essentially three-dimensional, an additional degree of freedom appears and φ_0 changes from the two-dimensional case.

To explain Dickinson and Long's 1978 data, Long (1978) introduced a quantity called the 'action' of the grid and proposed in his qualitative theory that the flow near the planar oscillating grid be modelled by a system of doublets of opposite sign placed regularly

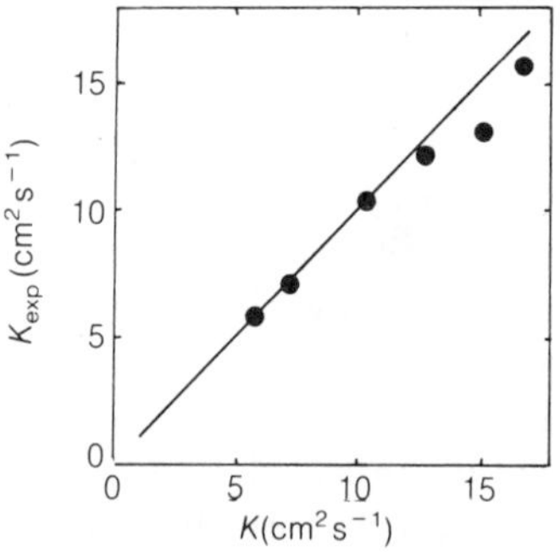

Figure 5.28 *Comparison of values of the grid action calculated from (5.5.11) (K) with values experimentally measured (K_{exp}) by Dickinson and Long (1978) in experiments with $f = 6.3–12.0$ Hz. The straight line corresponds to $K = K_{exp}$.*

in infinitesimal holes in a rigid plane. Long's theory does not allow us to determine the dependence of the action K on external parameters: it only predicts that $K \propto f$. The values of K must be determined from the experimental dependences

$$H = (Kt)^{1/2}.$$

Now, with the help of (5.4.5), (5.5.5) and (5.5.10), we can calculate Long's action

$$K = 2c_3 \beta v \, Re \tag{5.5.11}$$

and compare the calculated values with those (K_{exp}) obtained experimentally by Dickinson and Long (1978). This comparison is shown in Figure 5.28. Thus, Long's action acquires a clear physical sense: at the limit $Re \to \infty$ ($\beta \to$ const, see Figure 5.18) the action is equal, in fact, to the forcing amplitude $K \simeq Q/v$. This gives $K \propto f^{3/2}$, in agreement with Dickinson and Long's experimental data.

Finally, note that despite the good agreement shown in Figure 5.28, one important question remains: is φ_0 for $\tau > 1$ for the three-dimensional case a universal constant that does not depend on f, d and ε, as in the two-dimensional case?

References

Afanasyev, Ya.D. and Voropayev, S.I. (1991) Plane vortex flow induced by a mass source/sink in a rotating viscous fluid. *Izv. Acad. Nauk SSSR, Mekh. Zhid. i Gaza,* **4**, 172–5.

Afanasyev Ya.D., Voropayev S.I. and Filippov, I.A. (1988) Laboratory investigation of flat vortex structures in a stratified fluid. *Dokl. Akad. Nauk SSSR*, **300**, 704–7.

Afanasyev, Ya.D., Voropayev, S.I. and Filippov, I.A. (1990) Conductivity microprobe for fine structure measurements in a stratified fluid. *Okeanologiya*, **30**, 502–4.

Ahlnas K., Royer, T.C. and George, T.H. (1987) Multiple dipole eddies in the Alaska coastal current. *J. Geophys. Res.*, **92**, 41–7.

Arfken, G. (1985) *Mathematical Methods for Physicists*, 3rd edn. Academic Press, New York.

Baker, D.T. (1966) A technique for the precise measurements of small fluid velocities. *J. Fluid Mech.*, **26**, 573–5.

Barenblatt, G.I. (1979) *Similarity, Self-Similarity, and Intermediate Asymptotics.* New York.

Barenblatt, G.I. and Voropayev, S.I. (1983) A contribution to the theory of steady turbulent layers. *Izv. Akad. Nauk SSSR, Fiz. Atmos. Okeana*, **19**, 169–74.

Barenblatt, G.I., Voropayev, S.I. and Filippov, I.A. (1989) Model of Fedorovian coherent structures in the upper ocean. *Dokl. Akad. Nauk SSSR*, **307**, 720–4.

Batchelor, G.K. (1967) *An Introduction to Fluid Dynamics.* Cambridge University Press.

Benilov, Y.A., Voropayev, S.I. and Zhmur, V.V. (1983) Simulation of the evolution of the upper turbulent ocean layer during heating. *Izv. Akad. Nauk SSSR, Fiz. Atmos. Okeana*, **19**, 175–84.

Birkhoff, G. (1960) *Hydrodynamics, a Study in Logic, Fact and Similitude*, 2nd edn. Princeton University Press.

Cantwell, B.J. (1978) Similarity transformations for the two-dimensional, unsteady, stream-function equation. *J. Fluid Mech.*, **85**, 257–71.

Cantwell, B.J. (1986) Viscous starting jets. *J. Fluid Mech.*, **173**, 159–89.

Cantwell, B. and Rott, N. (1988) The decay of a viscous vortex pair. *Phys. Fluids*, **A31**, 3213–24.

Carslow, H.S. and Jaeger, J.C. (1960) *The Conduction of Heat in Solids*, 2nd edn. Clarendon Press, Oxford.

Couder, Y. and Basdevant, C. (1986) Experimental and numerical study of vortex couples in two-dimensional flows. *J. Fluid Mech.*, **173**, 225–51.

Cromwell, T. (1960) Pycnoclines created by mixing in an aquarium tank. *J. Mar. Res.*, **18**, 72–82.

Dickinson, S.C. and Long, R.R. (1978) Laboratory study of the growth of a turbulent layer of fluid. *Phys. Fluids*, **21**, 1698–701.

Drake, Ch., Imbrie, J., Knauss, J. and Turekian, K. (1978) *Oceanography*. Holt, Rinehart and Winston, New York.

Dritschel, D.G. (1986) The nonlinear evolution of a rotating configuration of uniform vorticity. *J. Fluid Mech.*, **172**, 157–82.

Dwight, H.B. (1961) *Tables of Integrals and Other Mathematical Data*. Macmillan, New York.

Falco, R.E. (1977) Coherent motions in the outer region of turbulent boundary layers. *Phys. Fluids*, **20**, 124–32.

Fedorov, K.N. and Ginzburg, A.I. (1988). *The Surface Ocean Layer*. Gidrometeoizdat, Leningrad.

Fernando, H.J.S. (1991) Turbulent mixing in a stratified fluid. *Annu. Rev. Fluid Mech.*, **23**, 455–93.

Flierl, G.R., Stern, M.E. and Whitehead, J.A. (1983) The physical significance of modons: laboratory experiments and general integral constraints. *Dyn. Atoms. Oceans*, **7**, 233–63.

Gibson, C.H. and Schwarz, V.H. (1963) Detection of conductivity fluctuations in a turbulent flow field. *J. Fluid Mech.*, **27**, 848–55.

Ginzburg, A.I. and Fedorov, K.N. (1984) The evolution of a mushroom formed current in the ocean. *Dokl. Akad. Nauk SSSR*, **274**, 481–4.

Greenspan, H. (1968) *The Theory of Rotating Fluids*. Cambridge University Press.

Griffiths, R.W. and Linden, P.F. (1981) The stability of vortices in a rotating, stratified fluid. *J. Fluid Mech.*, **105**, 283–316.

Hopfinger, E.J. and Linden, P.F. (1982) Formation of thermocline in zero-mean-shear turbulence subjected to a stabilizing buoyancy flux. *J. Fluid. Mech.*, **114**, 157–73.

Hopfinger, E.J. and Toly, J.A. (1976) Spatially decaying turbulence and its relation to mixing across density interfaces. *J. Fluid Mech.*, **78**, 155–75.

Jackson, J.D. (1975) *Classical Electrodynamics*, 2nd edn. Wiley, New York.

Kamenkovich, V.M. (1973) *The Fundamentals of Ocean Dynamics.* Gidrometeoizdat, Leningrad.

Kamenkovich, V.M., Koshlyakov, M.N. and Monin, A.S. (1987) *Synoptic Eddies in the Ocean.* Gidrometeoizdat, Leningrad.

Kamke, E. von (1959) *Differentialgleichungen. Losungsmethoden und Losungen.* Leipzig.

Kantha, L.H. and Long, R.R. (1980) Turbulent mixing with stabilizing surface buoyancy flux. *Phys. Fluids*, **23**, 2142–3.

Kiehn, R.M. (1990) Topological torsion, Pfaff dimensions and coherent structures. In *Topological Fluid Mechanics* (Eds H.K. Moffat and A. Tsinober). Cambridge University Press.

Kiehn, R.M. (1992) Continuous topological evolution. *Phys. Rev.* **B** (submitted).

Kloosterziel, R.C. and van Heijst, G.J.F. (1991) An experimental study of unstable barotropic vortices in a rotating fluid. *J. Fluid Mech.*, **223**, 1–24.

Kochin, N.E., Kibel, I.A. and Rose, N.V. (1963) *Theoretical Hydromechanics*, Vol. 2, 3rd edn. Fizmatgiz, Moscow.

Korn, G.A. and Korn, T.M. (1961) *Mathematical Handbook for Scientists and Engineers. Definitions, Theorems and Formulas for Reference and Review.* McGraw-Hill, New York.

Kraus, E.B. and Turner, J.S. (1967) A one-dimensional model of the seasonal thermocline. The general theory and its consequences. *Tellus*, **19**, 92–106.

Kraus, E.R. (1977) *Modeling and Prediction of the Upper Layers of the Ocean.* Pergamon Press, Oxford.

Lamb, H. (1932) *Hydrodynamics*, 6th edn. Cambridge University Press.

Landau, L.D. (1944) New exact solution of the Navier–Stokes equations. *Dokl. Akad. Nauk. SSSR*, **44**, 311–14.

Landau, L.D. and Lifshitz, E.M. (1980) *Statistical Physics*, Part 1, 3rd edn. Pergamon, Oxford.

Landau, L.D. and Lifshitz, E.M. (1987) *Fluid Mechanics*, 2nd edn. Pergamon, Oxford.

Lee, T.D. (1962) *Mathematical Methods of Physics. A Course of Lectures Given at Columbia University*. Columbia University Press, New York.

Linden, P.F. (1975) The deepening of mixed layer in a stratified fluid. *J. Fluid Mech.*, **71**, 385–405.

Long, R.R. (1978) Theory of turbulence in a homogeneous fluid induced by an oscillating grid. *Phys. Fluids*, **21**, 1887–8.

McWilliams, J.C. and Zabusky, N.J. (1982) Interactions of isolated vortices. I: Modons colliding with modons. *Geophys. Astrophys. Fluid Dyn.*, **19**, 207–27.

Manakov, S.V. and Schur, I.N. (1983) Stochasticity in two-particles dispersion. *Pis'ma Zh. Tekh. Fiz.*, **37**, 45–8.

Martin, S., Simmons, W. and Wunsch, C. (1972) The excitation of resonant triads by single internal waves. *J. Fluid Mech.*, **53**, 17–44.

Massel, S.R. (1989) *Hydrodynamics of the Coastal Zone*. Elsevier, Amsterdam.

Merzkirch, W. (1974) *Flow Visualization*. Academic Press, New York.

Monin, A.S. and Ozmidov, R.V. (1981) *Oceanic Turbulence*. Gidrometeoizdat, Leningrad.

Morton, B.R. (1959) Forced plumes. *J. Fluid Mech.*, **5**, 151–63.

Munk, W.H., Scully-Power, P. and Zachariasen, F. (1987) Ships from Space (The Bakerian Lecture). *Proc. R. Soc. Lond.* **A412,** 231–59.

Nekrasov, A.I. (1931) Diffusion of a vortex. *Papers of TsAGI*, **84**, 1–32.

Nguyen Duc, J.M. and Sommeria, J. (1988) Experimental characterization of steady two-dimensional vortex couples. *J. Fluid Mech.*, **192**, 175–92.

Orlandi, P. (1990) Vortex dipole rebound from a wall. *Phys. Fluids*, **A2**, 1429–36.

Orlandi, P. and van Heijst, G.J.F. (1992) Numerical simulation of tripolar vortices in 2D flow. *Fluid Dyn. Res.*, **9**, 179–206.

Oster, G. (1965) Density gradients. *Sci. Am.*, **213**(1).

Robinson, A.R. (ed.) (1983) *Eddies in Marine Science*. Springer-Verlag, Berlin.

Robinson, S.K. (1991) Coherent motions in the turbulent boundary layer. *Annu. Rev. Fluid. Mech.*, **23**, 601–39.

Rouse, H. and Dodu, J. (1955) Diffusion turbulence à travers une discontinuitè. *Houille Blanche*, **10**, 405–10.

Schlichting, H. (1933) Laminare strahlausbreitung. *Z. Angew. Math., Mech.*, **13**, 260–3.

Schlichting, H. (1979) *Boundary Layer Theory*, 7th edn. McGraw-Hill, New York.

Sedov, L.I. (1959) *Similarity and Dimensional Methods in Mechanics*. Academic Press, New York.

Sheres, D. and Kenyon, K.E. (1989) A double vortex along the California coast. *J. Geophys. Res.*, **94**, 4989–97.

Simpson, J.E. (1987) *Gravity Currents in the Environment and the Laboratory*. Ellis Horwood, Chichester.

Slezkin, N.A. (1934) On the case when the equations of motion for viscous fluid can be integrated. *Sci. Papers Moscow State University*, **2**, 89–90.

Spiegel, E.A. and Veronis, G. (1960) On the Boussinesq approximation for a compressible fluid. *Astrophys. J.*, **131**, 442–47.

Squire, H.B. (1951) The round laminar jet. *Q. J. Mech. Appl. Maths*, **4**, 321–9.

Stern, M.E. (1987) Horizontal entrainment and detrainment in large-scale eddies. *J. Phys. Oceanogr.*, **17**, 1688–95.

Stern, M.E. and Voropayev, S.I. (1984) Formation of vorticity fronts in shear flow. *Phys. Fluids*, **27**, 845–55.

Stockton, R.L. and Lutjeharms, J.R.E. (1988) Observations of vortex dipoles on the Benguela upwelling front. *SA Geographer*, **15**, (1/2), September 1987/April 1988.

Thompson, S.M. and Turner, J.S. (1975) Mixing across an interface due to turbulence generated by an oscillating grid. *J. Fluid Mech.*, **67**, 349–68.

Turner, J.S. (1964) The flow into an expanding spherical vortex. *J. Fluid Mech.*, **18**, 195–208.

Turner, J.S. (1966) Jets and plumes with negative or reversing buoyancy. *J. Fluid Mech.*, **26**, 779–92.

Turner, J.S. (1968) The influence of molecular diffusion on turbulent entrainment across a density interface. *J. Fluid. Mech.*, **33**, 638–56.

Turner, J.S. (1969) Buoyant plumes and thermals. *Annu. Rev. Fluid Mech.*, **1**, 29–44.

Turner, J.S. (1973) *Buoyancy Effects in Fluids.* Cambridge Univeristy Press.

Van Dyke, M. (1975) *Perturbation Methods in Fluid Mechanics*, 2nd edn. Parabolic Press, Stanford.

Van Dyke, M. (1982) *An Album of Fluid Motion.* Parabolic Press, Stanford.

van Heijst, G.J.F. and Flor, J.B. (1989) Dipole formation and collisions in a stratified fluid. *Nature*, **340**, 212–15.

van Heijst, G.J.F. and Kloosterziel, R.C. (1989) Tripolar vortices in a rotating fluid. *Nature*, **338**, 569–71.

Vastano, A.C. and Bernstein, R.L. (1984) Mesoscale features along the first Oyashio intrusion. *J. Geophys. Res.* **89**, 587–96.

Voropayev, S.I. (1983) Free jet and frontogenesis in shear flow. *Tech. Rep. Woods Hole Oceanographic Institute*, WHOI-83-41, 147–59.

Voropayev, S.I. and Afanasyev, Ya.D. (1991) Steady horizontal jet in a stratified fluid. *Mech. Res. Commun.*, **18**, 435–40.

Voropayev, S.I. and Afanasyev, Ya.D. (1992) Two-dimensional vortex dipole interactions in a stratified fluid. *J. Fluid Mech.*, **236**, 665–89.

Voropayev, S.I. and Afanasyev, Ya.D. (1993a) Symmetric interaction of developing horizontal jet in a stratified fluid with a vertical cylinder. *Phys. Fluids* (submitted).

Voropayev, S.I. and Afanasyev, Ya.D. (1993b) Vortex quadrupoles and two-dimensional oscillating grid turbulence. *Appl. Sci. Res.* **51**, 475–80.

Voropayev, S.I. and Filippov, I.A. (1985) Development of a horizontal jet in homogeneous and stratified fluids: laboratory experiments. *Izv. Akad. Nauk SSSR, Fiz. Atmos. Okeana*, **21**, 964–72.

Voropayev, S.I. and Neelov, I.A. (1991) Laboratory and numerical modeling of vortex dipoles (mushroom-like currents) in a stratified fluid. *Okeanologiya*, **31**, 68–75.

Voropayev, S.I., Afanasyev, Ya.D. and Filippov, I.A. (1991) Horizontal jets and vortex dipoles in a stratified fluid. *J. Fluid Mech.*, **227**, 543–66.

Voropayev, S.I., Afanasyev, Ya.D. and van Heijst, G.J.F. (1993)

Experiments on the evolution of gravitational instability of an overturned, initially stably stratified fluid. *Phys. Fluids* (in press).

Voropayev, S.I., Afanasyev, Y.D. and van Heijst, G.J.F. (1993a) Two-dimensional flows with zero net momentum: evolution of vortex quadrupoles and propagation of turbulent region. *J. Fluid Mech.*, in press.

Voropayev, S.I., Gavrilin, B.L., Zatsepin, A.G. and Fedorov, K.N. (1980) A laboratory study of the deepening of a mixed layer in a homogeneous fluid. *Izv. Akad. Nauk SSSR, Fiz. Atmos. Okeana*, **16**, 197–200.

Woods, J.O. (1980) Do waves limit turbulent diffusion in the ocean? *Nature*, **288**, 219–24.

Index